CW00525658

The Search for Extraterrestrial Life

The Search for Extraterrestrial Life

Essays on Science and Technology

Edited by
P. DAY

Oxford   New York   Tokyo
OXFORD UNIVERSITY PRESS
THE ROYAL INSTITUTION
1998

Oxford University Press, Great Clarendon Street, Oxford OX2 6DP

Oxford   New York
Athens   Auckland   Bangkok   Bogota   Bombay
Buenos Aires   Calcutta   Cape Town   Dar es Salaam
Delhi   Florence   Hong Kong   Istanbul   Karachi
Kuala Lumpur   Madras   Madrid   Melbourne
Mexico City   Nairobi   Paris   Singapore
Taipei   Tokyo   Toronto   Warsaw

and associated companies in
Berlin   Ibadan

Oxford is a trade mark of Oxford University Press

Published in the United States
by Oxford University Press Inc., New York

A catalogue record for this book is available from the British Library

Library of Congress Cataloging in Publication Data
(Data applied for)

ISBN 0 19 850414 4

Typeset by EXPO Holdings, Malaysia

Printed in Great Britain by
Bookcraft Ltd., Midsomer Norton, Avon

PREFACE

This annual volume, prepared for the Members of the Royal Institution, brings together the texts of Friday Evening Discourses delivered over the last year of so in the historic Lecture Theatre at 21 Albemarle Street. As in earlier volumes, the topics covered range very widely over many aspects of contemporary science and technology. In addition, following the tradition of the Royal Institution, others also treat historical or biographical subjects relating to science and its impact on society. It is to be hoped that this collection gives an impression of the continuing vitality of Friday evenings at the Royal Institution, as a medium through which scientists and engineers expound their work to lively general audiences.

As Editor, my warm thanks are due to Sarah Cripps and Evie Jamieson for keeping track of the manuscripts and the publishing process, and finally, of course, to the authors for putting their Discourses into printed words.

London   P.D.
May 1998

CONTENTS

PLATES

The plates section falls between pages 68 and 69.

1. The sun in both visible light (*right*) and in X-rays (*left*). The dark areas visible (sunspots) are seen to be the seat of energy which manifests itself as X-rays in the solar corona. (Yokoh satellite.)
2. Patches of orange carbonates on the surface of martian meteorite ALH 84001. The field of view is about 0.5 mm across. The grains were produced below the surface of Mars when water circulated through it.
3. The historiated letter 'I' from *Genesis* in the Lucka bible.
4. Magnified ($\times 1000$) portion of the dark grey column depicted in a sixteenth century German manuscript showing the presence of at least seven pigments.
5. Photographs and microphotographs of elaborately decorated initials on a fifteenth century German manuscript MS Ger 4 (f 2r and f 28v) from the DMS Watson library.
6. Faces on the Byzantine/Syriac lectionary (upper f 188v, lower f 67v) blackened by degradation of white lead to black lead sulphide.
7. Partially reconstructed glazed bowl painted in black on a blue background (CF 1412). The scale section on the left is 2.5 cm long.
8. Polychromatic Egyptian faience samples from the Petrie Museum, (a) UC. 686 red pigment identified to be red ochre and (b) UC. 888 (lotus) yellow pigment identified to be lead(II) antimonate, $Pb_2Sb_2O_7$ (scale in mm).

CONTRIBUTORS

John Burland
Professor of Soil Mechanics,
Department of Civil Engineering,
Imperial College,
London SW7 2BU

Robin J.H. Clark
Professor of Inorganic Chemistry,
University College,
20 Gordon Street,
London WC1H 0AJ

Peter Day
Director and Fullerian Professor
   of Chemistry,
The Royal Institution of
   Great Britain,
21 Albermarle Street,
London W1X 4BS

Monica M. Grady
Department of Mineralogy,
The Natural History Museum,
Cromwell Road,
London SW7 5BD

Dan McKenzie
Department of Earth Science,
University of Cambridge,
Madingley Road,
Cambridge CBJ 0EZ

Andrew Wallard
Deputy Director, National
Physical Laboratory,
Teddington,
Middlesex TW11 0LW

Sir Arnold Wolfendale
Department of Physics,
University of Durham,
South Road,
Durham DH1 3LE

Will Wyatt
Chief Executive,
BBC Broadcast,
Broadcasting House,
Portland Place,
London W1A 1 AA

The search for extraterrestrial life—and the future of life on Earth

ARNOLD WOLFENDALE

Introduction

One of the great questions for humanity is 'Are we alone?' The question is not a new one, of course, but has come to the fore over the last year or two for a number of reasons. First, radio astronomy has achieved such high sensitivities that to look for signs of life from the directions of other stars, by way of 'intelligent' radio signals, is becoming practicable. Secondly, recent claims for the detection of past (and present?) 'life' on Mars has raised the possibility of the actual formation of 'elementary' life being very common. Finally, the observations of 'Earth-grazing' comets and asteroids has pointed out the threat to life on Earth from the impacts of these bodies. Similar impacts on other (potential) life-bearing planets is one of the factors which must be considered when estimating how many 'civilizations' there are likely to be in the Galaxy.

Historical attitudes

The idea, commonly held nowadays, that 'intelligent life' is quite common in the Galaxy is not a new one. From many examples, one can quote just two: by the Greek Philosopher, Lucretius and by the Chinese Philosopher, Teng Mu.

> The universe is infinitely wide. It's vastness holds innumerable atoms so it must be unthinkable that our sky and our round world are precious and unique ......
> Out beyond our world there are, elsewhere, other assemblages of matter making other worlds.
> Ours is not the only one in airs embrace.
>
> Lucretius First century BC
(after Drake and Sobel[1])
>
> How unreasonable it would be to suppose that, besides the heaven and earth which we can see there are no other heavens and no other earths.
>
> Teng Mu AD Thirteenth century[1]

However, by the sixteenth and seventeenth centuries such views were anathema to the Christian Church and highly dangerous ones to hold. The Church's teaching was, essentially, that the Universe was perfect, with the exception of the earth on which resided those unique, generally sinful, human beings whose redemption could only be arranged via the Church's intervention. It was natural to teach, therefore, that the Earth was at the centre of all things and woe betide anyone who thought otherwise. As is well known, Giordano Bruno (1548–1600), the Italian monk, paid the price for a twofold heresey: that it was the sun, not the earth, that was at 'the centre', and that we are not alone. He was burned at the stake in 1600.

> 'Innumerable suns exist; innumerable earths revolve about these suns in a manner similar to the way the seven planets revolve around our sun. Living beings inhabit these worlds'.
>
> Giordano Bruno Sixteenth century[1]

Bruno had been influenced by Nicholas Copernicus (1473–1543) whose famous book in the year of his death put forward the idea of a sun-centred planetary system.

It was Galileo Galilei (1564–1642) who really clinched the idea, thereby demoting the Earth from its pride of place. This brilliant physicist-cum-astronomer used the newly invented telescope to observe the heavens and discovered the now-called Galilean satellites orbiting Jupiter (Fig. 1). Galileo's view was that since the Moon orbits the Earth and the satellites orbit Jupiter so will the planets orbit the Sun—their

Fig. 1 Jupiter and a satellite—a modern Hubble Space Telescope photograph. Galileo would also have seen the satellites away from the planet's disc; he observed that they were rotating round the planet.

nearest big mass. Galileo, too, received the attention of the Inquisition and was lucky to escape with his life (see, for example, Sharrat[2]); instead of death he suffered a form of house-arrest.

In many ways, Galileo was the first mathematical physicist and his work on inertia, falling bodies, the orbits of projectiles and on pendulums, was ground-breaking; this work was built on by Isaac Newton, who was born the year that Galileo died.

The Drake equation

Having established that the Earth is not at the centre of everything and knowing, as we do, that the sun appears to be a rather typical star, the stage is set for a serious look at the possibilities for life elsewhere.

It can be said, immediately, that there is, as yet, no evidence at all for intelligent life elsewhere. Serious attempts are being made with large radiotelescopes (such is that at Arecibo, Fig. 2) to detect intelligent

Fig. 2 The Arecibo radio telescope in Puerto Rico. This telescope is used from time to time to search for radio signals from extraterrestrial intelligent life.

signals but without success so far; this SETI project (see NASA brochure, 1990;[3] Project Phoenix, 1994[4]) was funded initially by NASA but after support was withdrawn the research groups have had to rely on private donations. The present null results show that there is no detectable intelligent life out to a few tens of light years from the Sun— not a negligible distance, but one small compared with the dimensions of the Galaxy (distance to the Galactic Centre $\simeq$ 25 000 light years).

Because of the lack of evidence we are reduced to hypothesis and thus to an estimate of how probable extraterrestrial (ET) intelligent life is, i.e. how many such locations there might be. It is usual, in this context, to use the so-called Drake equation, named after Frank Drake, the American scientist, who has done so much work in this field. Before giving the equation it is important to stress just what it is that it gives the rule for estimating. It relates to the number of *detectable* civilizations and it assumes that the intelligent life is *similar* to ours in that it resides on a planet orbiting a star. It also assumes that the message is recognizable as such. Thus life-forms very different from ours are not included (we have enough problems trying to work out the numbers for earth-types without spreading the net wider!).

The Drake equation runs as follows:

The Drake equation

$$N = Rf_1\, nf_2f_3f_4T$$
R = rate of star formation
f_1 = fraction of stars with planets
n = number of planets hospitable to life per star
f_2 = fraction where life emerges
f_3 = fraction where intelligent life appears
f_4 = fraction capable of communication
T = length of time for which life remains detectable

Although one can make criticisms of the equation from the standpoint of the interdependence of some of the terms, and so on, it does give a list of the obviously relevant parameters. It is clear that, although some of the parameters are known, others are not, and this relates primarily to the biological ones. As an astrophysicist, however, I will have to concentrate mainly on the astronomical factors, although the recent work on the Martian meteorite will also receive attention.

Before continuing, it is necessary to make the rather obvious remark that the uncertainties in our knowledge of many of the parameters are so great that no credence can, yet, be put on a derived value for 'the number of ...'. Nevertheless, progress is being made all the time with the individual terms and it is useful to make an assessment. The value of 'T', the length of time for which intelligent life remains detectable, is one that will receive particular attention insofar as it has great relevance to the human condition, namely the 'future of life on earth'.

Life on Mars?

It has long been the stuff of science fiction that there is life on Mars—'little green men' and that sort of thing. Although in recent times such ideas have been dismissed as fanciful, the view that a very primitive form of life may have existed at very early epochs (say some 4 billion years ago), when the necessary free water was available, has remained respectable. The idea took something of a knock when the American Viking Orbiter/Landers (1976–82) sent back some 50 000 photographs of the Martian surface, and other data, which gave no signs of life of any form. The Martian surface appears to be a most inhospitable place.

Potential help was at hand, however, with the demonstration that a certain class of meteorites found on earth were, in fact, of Martian origin. Excitement reached fever pitch with the claim by NASA that one of these meteorites contained evidence of 'life' in the form of fossils. Figure 3 summarizes the life history of the rock chipped out of Mars by the impact of an asteroid some 16 million years ago. The techniques involved in determining the chronology involve the use of radioactive nuclei, some from the initial supernova, others induced by cosmic rays at different times. Understandably, the media reacted with great interest and much merriment. One of the ensuing newspaper cartoons is illustrated in Fig. 4. The more sober-sided scientific community reacted with great caution, the role of dramatic claims in releasing much-needed research funds not being lost on them! The scientific problems associated with the discovery were—and still are—mainly associated with the claimed fossils—they are almost entirely much smaller than those on earth—and there is a lack of evidence for cell-walls. Figure 5 shows the biggest of the small fossils claimed by the NASA group.

It is appropriate to point out that some meteorites have been known for some time to contain quite complicated biological-style molecules

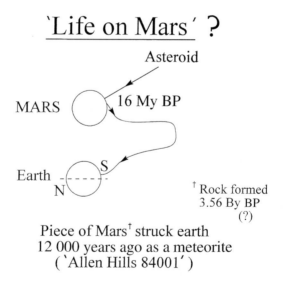

'Life on Mars' ?

Asteroid

MARS   16 My BP

Earth   S
      N

[†] Rock formed
3.56 By BP
(?)

Piece of Mars[†] struck earth
12 000 years ago as a meteorite
('Allen Hills 84001')

Research group (NASA) [OU, Nat. History Museum]

claims to have found primitive fossils....

Fig. 3 Life history of the Martian meteorite 'Allen Hills 84001'.

'Okay. It's only circumstantial evidence but let's take a vote. Hands up those who think there's life on Mars . . .'

Fig. 4 Cartoon from the *Daily Mail*, 8 August 1996.

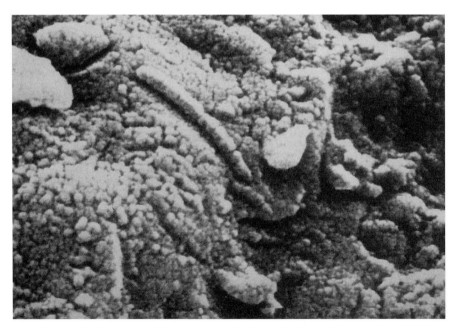

Fig. 5 An example of the sort of structure claimed by the NASA group in 1996 to be a fossil.

(amino acids); the problem has been the lack, until now(?), of fossils of true replicating systems, namely systems that could, in principle, join the life-sequence.

The jury is still out here, but it is interesting, and important, to note that recent work at The Open University and the Natural History Museum (described at a recent meeting at The Royal Society) has provided evidence by way of isotopic ratios in Martian meteoritic material which gives support to the idea of very primitive life-forms. We must conclude, then, that one of the important terms in the Drake equation ('f'_2—the fraction where life emerges) *may* be quite appreciable.

The search for planets round other stars

Staying with the Drake equation we turn our attention to f_1, the fraction of stars with planets. It might have been thought that this number was known with reasonable accuracy but this is not the case. The parent stars are so bright compared with their (possible) planets that the latter cannot be seen directly. It is true that accretion disks of solid material have been seen round a number of stars—with some nice recent examples from the Hubble Space Telescope—and these are generally regarded as the source of planets-to-be—but that is not the same thing as detecting actual planets.

Help is at hand, however, from neutron stars, the incredibly condensed stars which form when a more normal star collapses. Essentially, the protons in the hydrogen atoms eat up the electrons to form neutrons and neutrinos, the latter, having phenomenal penetration, immediately escape. The stars thus contracts by a factor of about 100 000—this being the ratio of atomic to nuclear dimensions—and the gentle rotation of the original star speeds up accordingly. For some reason, not yet fully understood, the neutron star has a magnetic field which is not aligned with the axis of rotation and this rotating field causes effects in the surrounding plasma which generate a rotating beam of radio waves (and other forms of electromagnetic radiation and fast atomic particles). The result is the famous 'pulsar', or pulsating radio star, discovered in 1967 by Jocelyn Bell and Tony Hewish.

The pulsars have, understandably, remarkably good time-keeping properties and it is this feature which has allowed extra-solar system planets to be inferred. If we imagine two skaters on an ice rink joined together by a taut rope going round one another in circles then, if one is much more massive than the other, the lighter one will go around in a much bigger circle than the heavy one. Replace the heavy skater by a pulsar, the light one by a planet and the rope by their gravitational attraction and we have a good analogy. In fact, because the ratio of the masses is very great ($\approx 10^6$) the orbit of the pulsar is very small but it is just sufficient for the radio beam to be 'modulated' in frequency because of the motion of the source and for this modulation to be detected.

The present situation is that one pulsar (PSR 1257 +12) with a millisecond period has at least three planets circling it, with masses from 0.015 to 2.8 times the mass of the Earth. There are also claims for three other pulsars having associated pulsars.

Now these planets will not be inhabited—the radiation level from the pulsar's emissions will be intolerable—but at least *they are there.* Indeed, one can argue that if pulsars can have planets then stars of a much gentler disposition should also have them—the point being that pulsars are born in the most dramatic circumstances following stellar collapse.

Turning to more normal stars the difficulties of detection are very great indeed but a start has been made at inferring the presence of planets much more massive than the Earth, specifically of mass similar to that of Jupiter (M_J). To date there have been seven good detections for stars out to 55 light years from the Sun, and five possible detections. The planets have masses in the range 0.5–5 M_J.

It would not surprise me at all if Drake's f_1 were quite high, say 0.01 or so.

T—the time for which life remains detectable

The parameter, *T*, is of course, one that has great relevance to the future of our own life as well as being of importance for life on other planets. Attention devoted to *T*, therefore, is time well spent whether there is intelligent life out there, or not!

Concerning the continuation of life on earth what are the mechanisms to be considered? First, there is the effect of humankind's own activities—the possible reduction of fertility, pollution of the environment, etc. This important aspect will not be considered here, again, we will stick with the astrophysics.

The astronomical hazards can be listed as: comets and asteroids, local stars exploding, and mischief with the Sun.

Starting with comets and asteroids—and large meteorites, too—their potential hazards for the earth are well known, although the actual frequency of impacts is rather uncertain. A fairly recent summary of likely rates is given in Box 4.

Likely rates of impact of 'bolides' (meteorites/asteroids/comets) with earth—and their effect.

(after Ahrens and Harris, 1992)[5]

Average time between impacts	Bolide diameter (km)	Effect
1000 y	0.1	Local destruction (e.g. Tunguska, 1908)
100 000 y	1	Would kill ~ 25% of the population
100 My	10	Would probably kill everyone

At first sight, we can put $T = 100$ million years (My) for life on Earth but there are some complications. For a start, the rates (or, more precisely, intervals, T) may be variable. Specifically, the rate of cometary impacts may increase over the next 20 000–30 000 years as we approach another star (α-Centaui). This star will, by gravitational attraction, so perturb the great Oort Cloud of Comets which extends out to nearly halfway to the next star, that more comets will flood into the Inner Solar System. Very recent work[6] has led to the claim that for other reasons, too, we might be in for an enhanced rate of impacts of cometary debris in forthcoming centuries. Indeed, it seems that the present period may be a comparatively quiet one from the standpoint of impacts of bodies of all sizes.

Returning to Box 4, we have a very rough and ready confirmation that $T \approx 100$ My for 10 km diameter comets from the fact that it is very likely indeed that the dinosaurs died out 65 My ago because of the impact of such a comet. The confirmation comes from a variety of sources, including the location—in the Gulf of Mexico—of the likely impact zone.

Other things being equal, it seems that T for the Earth could well be of order 100 My from the standpoint of the impact of large comets. However, help may be at hand, in view of the distinct possibility of firing a nuclear weapon at the comet and causing partial disruption and thereby a near-miss of the comet with the Earth, instead of a direct hit. The later such a civilization-affecting comet appears the more advanced will be our technology and thus the better the chance of negating its effect.

Allowing for all the above: increased comet rates, destruction possibilities and so on, it looks as though T is much bigger than 100 My for us.

What T might be for other planetary systems is, at present, anyone's guess. No doubt some planetary systems will have more comets and smaller T and others the reverse but we cannot yet be quantitative. It would surely be unlikely if all systems had T much less than ours, although it may be that possession of a large planet, such as our Jupiter, is a prerequisite for 'large T' and most planetary system may not have one. Jupiter's role for us is very interesting—it deflects many comets that venture into the Inner Solar System, where we are located, out of the Solar System all together; thus, a comet does not often have repeated chances of hitting the earth.

T—the lifetime of the Sun, and other stars

As mentioned already, 'mischief with the sun' is potentially a life-threatening phenomenon. At first sight, the sun is benign but in fact there *are* excitements from time to time in the form of solar flares (Plate 1). It is well known to cosmic ray workers that these flares can give rise to quite significant fluxes of energetic particles at the earth. The late George Wdowczyk and I[7] some years ago put forward evidence favouring rather dramatic effects on earth—on a statistical basis—every hundred thousand years or so. Proof positive is hard to come by but it does seem that, although there may be *some* effect, it is unlikely that life on Earth will be terminated by such phenomena.

Cometary impact, or more likely, still, our own incompetence, is a better bet if it is the demise of civilization that is required!

Similar remarks probably apply to a significant extent to other stars for most of their lifetimes. 'Lifetime' is, indeed, the parameter which can be guaranteed to be the maximum possible value of T. The Sun was born about 4.5 By ago and is about half-way through it's life. After about 3 By it will start to swell, eventually engulfing the Earth and, clearly, we will have had to leave a long time before that happens. This point has great significance for the whole 'life elsewhere?' problem, as will be demonstrated later.

The situation with many other stars is very similar. Admittedly, some stars explode dramatically and without much warning (supernovae) but they are few—only about one per 30 years in the Galaxy, and the Sun is not one of these. Furthermore, as Wdowczyk and I showed, the chance of another star sufficiently near us to be civilization-threatening is about once per 10 By, i.e. longer than the remaining solar life.[7]

The astro terms in the Drake equation

Taking everything together, it would be no surprise if, for a few per cent of the stars in the Galaxy, there were planets not too different from the earth. Comets, nearby stellar explosions and such like might limit the average value of *T* to, say 1 By, i.e. about one-tenth the age of the Galaxy. If the Mars detection of 'life' is correct then one finds that there could easily be several million sites where life exits. The big questions now are: How much of it is 'intelligent'? What fraction can communicate?

Perhaps 'absence of evidence' *is* 'evidence of absence'

The main argument against intelligent life being common is the fact that we have not seen any. At first sight this is bad science but in fact there is a good reason for saying it, for the following reason. As remarked earlier, when one's star is about to run out of fuel and become a giant it is 'time to go'. Now, very many stars in the Galaxy have been and gone, so that intelligent life on surrounding planets will have left. Thus, there should be very many space travellers visiting or passing by. The degree of intelligence—or technical mastery—does not need to be much superior to ours for this to happen. Despite many claims for the detection of UFOs none has been substantiated and, in my view, there is no firm evidence for visitors at all.

Thus, we may, in fact 'be alone'. [At this stage in the lecture, the lecturer's arguments appeared to be negated by the sudden appearance of a Martian, suitably clad, and with lights flashing. However, in view of 'its' inability to answer an elementary question about the mean mass, lifetime, and decay scheme of the muon, the lecturer categorized the visitor as 'unintelligent life' and argued that his thesis was still valid.)]

References

1. Drake, F. and Sobel, D., 1992, *'Is anyone out there?'—The scientific search for extraterrestrial intelligence'*, Souvenir Press.
2. Sharratt, M., 1994, *Galileo—decisive innovator*, Cambridge University Press.
3. SETI, 1990, Jill Tarter *et al.*, *NASA*, Ames Research Center, SETI Office, Moffett Field, California 94035-1000, USA.
4. Project Phoenix, 1994, *SETI Institute*, 2035 Landings Drive, Mountain View, California 94043, USA.
5. Ahrens, T.J. and Harris, A.W., 1992, *Nature*, **360**, 429.
6. Bailey, M.E., 1995, *Vistas in Astronomy*, **39**, 647.
7. Wdowczyk, J. and Wolfendale, A.W., 1977, *Nature*, **268**, 510.

SIR ARNOLD WOLFENDALE

Born 1927 in Rugby but brought up in Lancashire. He was educated at Stretford Grammar School and Manchester University where he joined the staff under the legendary P.M.S. Blackett. Moving with his family to Durham University in 1956 he built up a group working in Cosmic Ray Physics and later led a transition of much of the Physics Department's work into Astrophysics and Astronomy. He was elected to the Royal Society in 1977, was President of the Royal Astronomical Society, 1981–83, Astronomer Royal from 1991–95 and President of the Institute of Physics from 1994–96. Sir Arnold Wolfendale was Knighted in 1995. He has received a number of Honorary Degrees and Medals and has been honoured by a number of foreign Academies and Institutions. Sir Arnold has been giving Public Lectures for many years; he recently chaired a Government Committee looking into the Public Understanding of Science and several of the committee's recommendations have already been taken up. He has published some 500 papers. He is still trying to find out the origin of the cosmic radiation and feels that he is now very close—a feeling that he has had before!

Magellan looks at Venus

DAN MCKENZIE

The beginning

Unlike astrophysicists, earth scientists have until recently had only one object on which to test their ideas and theories about how the Earth formed and evolved. Furthermore, and perhaps surprisingly, the large-scale structure and dynamics of the Earth is in some ways harder to model than is that of stars. The problem is that nucleosynthesis inside starts is a very energetic process, and is controlled by nuclear reactions that have been studied in exquisite detail in the laboratory. Some of the most successful models of complicated natural processes have been developed to account for the observed abundances of elements and isotopes. The pressure and temperature within stars are also principally controlled by the interactions between photons, electrons, and ions, which can be studied in the laboratory. The principles governing the structure of stars were therefore understood before we had any comparable understanding of the Earth's structure and dynamics, for which we have had to depend on images produced by sound, rather than light, waves. Although we now think we understand, at least in outline, the processes involved, most of our ideas have come from observing what happens, and then arguing that we can understand what we see using simple physical principles based on laboratory experiments. Although such an approach is more sensible than the speculations it has replaced, and is obviously a useful way to start, it is possible that completely different processes may be important on other planets, and perhaps also early in the Earth's history. We would like to be in the same position as those who work on stellar structure. They can calculate the evolutionary path of a star as it grows older and compare it with a variety of observations from stars of different ages. One of the reasons astrophysicists have such confidence in their models is that they have a huge number of stars that they can observe, whereas the geologist has only one Earth. Hence the importance of looking at other planets and understanding how they work. As I have spent much of my scientific life thinking about the Earth, starting with the ideas of plate tectonics to describe surface deformation and then studying how mantle convection can maintain these motions, I was interested in planets that were likely still to be tectonically active. From this point of view the Moon was not a good bet. Although it is not completely inactive, the few moonquakes that take place are very small and occur at depths of about 700 km. Furthermore, they appear to be produced by the tides that are generated in the solid Moon by the Earth, rather than being an expression of large-scale tectonic deformation. Mars is bigger and is more likely to be tectonically active. However, much of its surface is old and probably not now being deformed. It does, however, have the largest volcanoes yet found in the solar system. With some luck, the spacecraft that are planned over the next few years will allow us to understand more about its interior. The experiments carried out by the earlier US landers were dominated by the desire to find life, and therefore were not much help in studying geological problems. But Venus was always much the most promising planet from my point of view, and I jumped at the chance to study it when it came, in about 1980, in the form of a boring technical document in a cheap envelope entitled 'Announcement of Opportunity: Venus Orbiting Imaging Radar'. At that time NASA sent out hundreds of such documents (they are now on the web), and you could send in a proposal to work on the data, which I did. And so began the project that has lasted the longest of any that I have ever undertaken.

Before Magellan I had never had anything to do with space, and knew nothing about what was involved. I had not thought about the difference between putting a satellite into orbit round the Earth, which is easy, and into orbit round Venus, which is very difficult indeed, because you must communicate with a small object at a great distance from the Earth, which is travelling very fast, and make it go into orbit as it passes the planet. The spacecraft that was finally sent to Venus had a transmitter whose power was about 300 W, and which could transmit about 200 k bits a second from the other side of the Solar System. At the beginning, the spacecraft was going to carry many different experiments. I went to my first meeting, at the Jet Propulsion Laboratory (JPL) in Pasadena, and was astonished when I entered the lecture room to find about 300 other scientists, all of whom had experiments they wanted to fly or who wanted to work on the data the spacecraft would return. Over the next few days they explained what they wanted to do, and I had my first experience of the opaque language that surrounds every aspect of JPL. Everything is referred to by a string of capital letters. I slowly got used to TBD (to be determined), BIDRS (basic image data records), MIDRS (mosaiced image data records), and so on. I still need a sheet which tells me what the many hundreds of such terms mean before I can understand the simplest talk. However, one thing I had understood before I became involved, which was that everything anyone was going to do *had* to be done with an instrument on the spacecraft, and so it was essential to make sure that *your* instrument was kept on and not other peoples'. I am not a builder of instruments, and the data I really wanted was going to come from the carrier frequency of the communications link, which *had* to be on board. So I was quite relaxed about this process, once NASA had agreed that I should be part of the project. But the same was not true of most of the people in the room, who were at their most persuasive. 'The Project', as it was called, gave some limited support to help people design their instruments, and we all went home to return in a year's time. When we did so, we were told that the estimated cost of the spacecraft was now $100 million more than it was last year! After the same thing happened for several years in a row, and the price tag reached about a billion dollars, NASA announced that the spacecraft was now too expensive, and would have to be completely redesigned to make it cheaper. 'The Project' then did two astonishing things: it threw off almost all the experiments that had been planned, on which many people had spent years of their lives, and reduced the mission to the task of making images of the surface of Venus. 'The Project' also collected a large number of existing bits and pieces of previous spacecraft to construct a remarkable object out of what was essentially NASA's version of the 'might come in handy' box of parts that I keep under the stairs.

The fundamental feature of Venus that dominates any attempt to map its surface is that is is always covered by clouds. They are made of droplets of sulphuric acid, and they never disappear. So photographs show nothing of the planetary surface. The only way forward is to use the spacecraft to illuminate the surface, in a waveband where the sulphuric acid is transparent, and in which the spacecraft can produce enough power to get a clear return. The spacecraft therefore must have a radar transmitter and receiver, so that it can illuminate the surface itself. But it cannot fly close to the surface, because it would burn up in the atmosphere, and must be powered by converting sunlight to electricity using solar panels. These problems dominated the final design (Fig. 1), which consisted on a big radar dish and two big flat solar panels to generate the power. The other features you can see are a long tapered horn at one side of the big dish, which is the altimeter. It works by sending the radar pulse straight downwards, to measure the height of

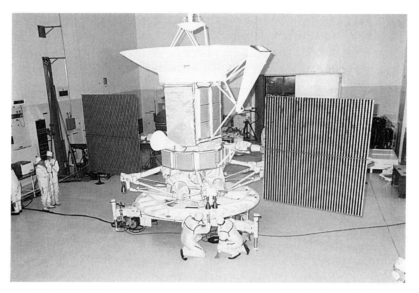

Fig. 1 The Magellan spacecraft being prepared for its launch. It is powered by electricity from solar panels, and uses the high gain antenna for mapping and communication with Earth. Rocket engine modules are used to control its orientation, and a star scanner to find its orientation.

the spacecraft above the surface and hence the shape of the planet. You can also see some small rocket motors on long arms sticking out from the bottom of the spacecraft. These are used to control its orientation. The main body of the spacecraft consists of a squat many-sided box, which contains the control bus, and a rectangular box beneath the big dish, which contains the radar. The big dish and the bus were spares left over from the Voyager mission, but the radar was new. The idea that saved a great deal of money was to use the *same* big dish to direct the radar on to a patch of the surface of the planet, to record the returned signals on tape recorders. Then the whole spacecraft rotates, to point the dish at the Earth, and the tape recorders play back the data, which are received by the Deep Space Network (needless to say, always known as the DSN). The DSN consists of three collections of antennae, in Owens Valley in California, near Madrid and near Canberra. This turn had to be very accurate, because the spacecraft had to point exactly at the Earth. And it had to do this twice every orbit, or twice every 3 hours, for at least 1 Venusian days of 243 Earth days. Rotating the spacecraft with rockets would use too much fuel, so it was done by spinning inertial wheels, in the same way as a cat can rotate to land on its feet when dropped from a high building. When I heard all this for the first time I thought 'Well, that's that. They will *never* get such a complicated system to work for so long'. I was wrong, but only just!

The next event was the Challenger explosion, which occurred after the spacecraft, by this time known as 'Magellan' (with, for some reason, a soft 'g') had been built. Like everyone, I was amazed, and horrified and fascinated by the stories that came out in the resulting enquiry. But I also had a personal interest in the outcome! The major effect on 'The Project' was that the large liquid hydrogen and oxygen rocket, that was to be used to propel Magellan out of the Earth's gravity field, had to be replaced by a solid fuelled rocket that was not so powerful. This in turn meant that there was less weight (and so less power) available for the rocket that was needed to insert Magellan into orbit round Venus, and the orbit was therefore going to be more elliptical. These problems were caused by NASA's unwillingness to fly liquid oxygen and hydrogen rockets in the Shuttle, but I could scarcely condemn them for their caution! So, finally, all was ready and the launch date was set, for the spring of 1989. As my teaching for the year was over, I thought I would go to Florida and watch the launch. I was astonished by the number of retired people who had come to watch the Shuttle in their campers. 'Oh yes', I was told, 'the number went up by 100 000 after Challenger: they are all here hoping for a repeat'. In fact I did not see the launch, because one of the hundreds of computers involved malfunctioned with 30 seconds to go. But, on 4 May 1989 Magellan was finally launched, pushed out of the Shuttle and sent on its way, with some grumbling from the astronauts about the dullness of this mission.

Thereafter, everything happened at JPL. Once the spacecraft was in orbit I had to get used to going there for 4 or 5 days every 2 months or so, which lasted for about 2 years. At the beginning I tried to change my body clock, but soon found that the best way was to go to bed at 5 in the afternoon, and go into JPL for the day at about 1 a.m. when the free-ways are empty and you can park. In some ways I was very glad when the project finally ended. I could go back to living quietly in a terrace house in Cambridge, and my major means of travel once again became my bicycle. But the troubles of 'The Project' were not yet over. Magellan arrived at Venus on 10 August 1990 and was successfully inserted into orbit when it was behind the planet. We all waited in great excitement at JPL for Magellan to reappear, going more slowly. When it duly did and at exactly the right speed, and a cheer went up. I did not under-stand why everyone was *so* happy, until Mars Observer failed, doing exactly the same manoeuvre. Everyone's contracts with NASA were conditional on successful orbit insertion. When Mars Observer failed, perhaps a hundred planetary scientists were left with 3 weeks to find another job, and the entire subject almost disappeared! After Mars Observer, I decided that people who worked with me on planetary geology would never be funded in this way, but would be poor but safe!

After orbit insertion was successful there was a huge party at a Country Club in the hills behind La Canada, which is where JPL *really* is: it only has a Pasadena address because of gerrymandering. Everyone involved in 'The Project' came, about 450 people altogether. The total cost of 'The Project' was about $1000 million, by *far* the most expensive experiment in which I had ever been involved. At the party I remember asking someone I was introduced to whether everything was now likely to be OK. 'No' she replied, 'I don't think the software has been properly tested: "The Project" is short of money and has cut corners'. I thought nothing more about this remark until the following week, when I was back in Cambridge and Magellan made the lunch-time news. All contact had been lost and the spacecraft was out of control, after having collected only a few, beautiful, test images. When I next returned to JPL, and after communications and control had been restored with Magellan, one of the people involved in the rescue explained how this type of disaster had been planned for, and how the spacecraft had been programmed to search for the Earth if it went out of control. This particular problem happened several times, but was not so alarming after the first event. It was finally traced to a software bug, using the exact copy of the spacecraft that had been built when it was manufactured, which had been kept on Earth.

To my astonishment Magellan worked perfectly for the first Venusian day. The only problem was that one of its tape recorders failed. Then, slowly, different parts of the electronics started to fail. For me this was the worst time. The spacecraft was returning wonderful images of the surface, almost everyone in 'The Project' was happy, NASA was giving press conferences and making noises about closing down 'The Project' to save money that it would spend on the space station, and the spacecraft was slowly dying. But I could not understand how most of the astonishing features on the planet had been produced, and knew that I needed the gravity field to have any chance of doing so. When the spacecraft was collecting images, it could not also be used to measure the gravity field. So I needed the imaging to stop. The big dish could then be pointed towards the Earth, not at Venus, when the spacecraft was nearest Venus. The line of sight component of the gravity field can then be obtained from the Doppler shift of the carrier frequency, by measuring its rate of change.

I was saved by the failure of part of the imaging system, which made it impossible to use Magellan to produce pictures of the surface. A different problem now arose, which was that so few people were interested in the gravity that NASA wanted to shut down 'The Project'. But a lot of letters and phone calls kept it alive for another 3 Venusian days measuring gravity, before NASA Headquarters finally ran out of patience and destroyed the spacecraft by burning it up in the atmosphere of Venus. They did so to prevent any further efforts by the gravity wallahs to keep it going 'for just *one* more day: please! It *only* costs $10 million a year'. And so the mission ended, having returned more data than all the previous deep space missions combined. Magellan must rank with Voyager as one of NASA's most successful deep space experiments ever. Those on 'The Project' felt that Magellan never got the attention it deserved, from either the press or from NASA Headquarters. Perhaps the problem was that it slowly built up the spectacular images, rather than returning a photograph in the glare of the television cameras.

But, scientifically, Magellan was an outstanding success. One small spacecraft had made images, at a resolution of about 150 m, of almost the entire surface of Venus. This is better coverage than we have of the Earth. Because 70% of our planet is covered by water, the sea floor must be imaged using sound waves, and many parts of the oceans have almost no coverage.

The altimeter measured the shape of almost the entire planet with a resolution of about 20 km, which again is better coverage than we have of Earth, because of the oceans. The gravity field that came from the Doppler tracking was also better determined than was that of the Earth 2 years ago, although the geophysical measurements have since improved. These are remarkable achievements, and the two people who I think are most responsible for this success are Steve Saunders, who was the project scientist at JPL, and Gordon Pettengill, a radar scientist at MIT who had total grasp of every aspect of the spacecraft. Steve has now moved to NASA Headquarters, and Gordon took early retirement in a hurry when Mars Observer failed. They are two of the most impressive scientists with whom I have worked. Much of the success of Magellan was due to their clear understanding of the scientific issues, and an ability to keep these in focus through the extraordinary logistic complications of managing a mission of this size. One of the great strengths of modern science is its independence of the individual. All of the major advances in which I have been involved would have been proposed by someone else a year or two later, and it is only with some effort that the community remembers to attach my name to these ideas for a few years, until they are either discarded or become so much part of the framework of the subject that no one bothers to say where they come from. This tendency is especially strong where, like Gordon and Steve, scientists are leaders of large teams. Yet, the success of projects, such as Magellan, depends entirely on such people devoting a large chunk of their lives to keeping the overall objectives clearly in mind throughout the whole process of the design and operation of the spacecraft, and its interaction with the DSN and JPL itself, whose activities for a year or two were dominated by the problem of dealing with the huge amounts of data that Magellan generated. I was entirely an onlooker throughout this operation, although I was included in all parts of it in exactly the same way as were my American colleagues who were US nationals working at US universities with NASA support. This lack of interest in where people work and originate is an aspect of US science that has always impressed me. Although I had expected that it would operate for Magellan also, I thought the stress on peoples' relationships as the key data became available might lead to some hint of nationalism, but it never did. This spirit has yet to arrive in Europe!

Synthetic aperture radar

The key idea underlying the Magellan Mission is that of synthetic aperture radar, or SAR. Ordinary radars on ships and aeroplanes work by pointing a radar dish in a known direction and measuring the time taken for radar waves to travel from the dish to the target and back. For good resolution they require big dishes and very large amounts of power, which is in short supply on a spacecraft. A SAR uses a different principle illustrated in Fig. 2, which depends on the Doppler effect rather than the size of the dish to obtain good resolution. On a SAR the dish is only used to focus the radar on a patch of the planet to one side of the ground track of the spacecraft. The frequency of echoes that come from parts of this patch that are exactly perpendicular to the orbit of the spacecraft are the same as that of the outgoing signal. However, that of echoes from behind the spacecraft are shifted to lower frequencies. Conversely, those from targets ahead of the spacecraft are shifted to higher frequencies. The distance of a target at right angles to the track can be found from the time taken for the echo to return. Because the SAR uses all the echoes that return to the spacecraft from a patch on the surface, it requires less power to map a planet than does a conventional radar, and can obtain good resolution with a small dish by using the Doppler effect. Although the radar images from the SAR look like photographs, they contain certain artefacts not present in ordinary images. One of the most trouble-

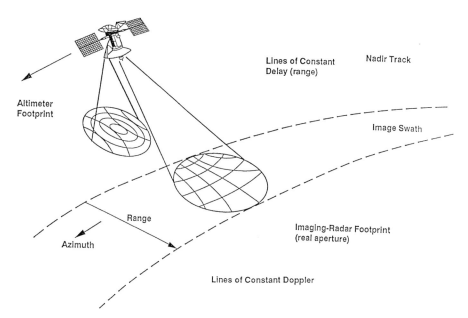

Fig. 2 How the SAR works. The nadir track is the path of the spacecraft on the surface of Venus. The SAR illuminates a patch of the surface (the footprint) to one side of the ground track, and uses the two way travel time of the echo (the delay) to measure the distance from the spacecraft, and its Doppler shift to measure its position along the strip that is being mapped (the image swath) (Ford 1993, *in* Ford *et al.* 1993).

some is called 'overlay', and arises because the radar return from the top of a steep hill arrives before that from its bottom. Figure 3(a) illustrates the problem, which causes the top of the hill to appear to be closer to the spacecraft than its base. Figure 3(b) shows a SAR image from a steep-sided glacial valley in Alaska. The glacier in the bottom of the valley is visible as a series of bright lines parallel to the sides of the valley. It is obscured in places by the returns from the top of the valley side towards the bottom of the image. In some areas of Venus this effect makes the images very difficult to interpret.

Figure 4 shows the sequence of operations that had to be carried out on every mapping orbit. Each such orbit mapped a strip about 25 km wide and 18 000 km long. Images like those reproduced here are built up by assembling many individual strips, and the dark lines on some show where there are data gaps. As this assembly continued, the extraordinary features on the surface of Venus were slowly revealed.

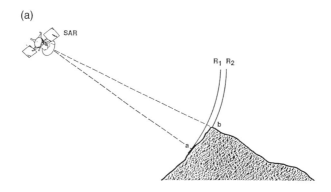

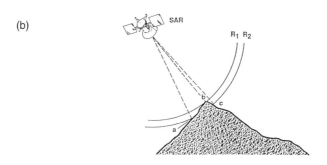

Fig. 3 (a) When a steep hill is illuminated at a shallow angle, the echo from the top, b, arrives after that from lower down, a. However when the illumination angle is steeper, (b), the reverse is true. The echo from the top of the hill then arrives first and overlies that from the base (Ford *et al.* 1989). (c) A SAR image from Alaska, showing a glacier at the bottom of a steep valley. In places the glacier is overlain by echoes from the valley side. Illumination is from the top (Farr 1993, *in* Ford *et al.* 1993).

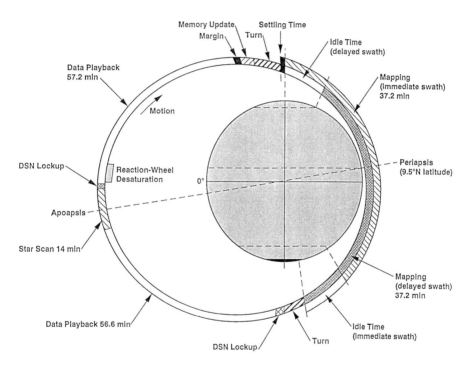

Fig. 4 The sequence of events that occurred during every 3-hour orbit of Magellan for 2 years (Ford 1993, *in* Ford *et al.* 1993).

Craters

All planets in the Solar System have been bombarded by meteorites which have produced craters, and Venus is no exception. Figure 5 shows an image of a medium sized crater, with a central peak, a flat floor with a circular crater wall, surrounded by lobate ejecta. The dark regions show areas that do not return radar energy, while the bright regions are strongly reflective. Whether or not a patch of the surface returns radar energy depends on its roughness at the radar wavelength of 12.6 cm, and is not related to its optical properties. The ejecta are radar-bright because they are rough, and contain boulders of all sizes. Some of these are sufficiently large to be visible as individual features on the SAR image shown in Fig. 5. Figure 6 is an image of the largest crater on Venus, known as Mead, with a diameter of 280 km. Like some large craters on the Moon, it has more than one ring. The craters on Venus differ in two ways from those on the Moon, Mars, and Mercury. There are few small (< 35 km diameter) craters, because small meteorites break up as they pass through the dense atmosphere: the pressure at the surface of Venus is 90 times greater than that on Earth. When meteorites break up in this way they deposit thick layers of dust like those visible as dark splotches in Fig. 7. The other difference is the ejecta, which flows out as lobes because it is fluidized by gas when the dense atmosphere is compressed by the impacting meteorite. These lobes are clear in Fig. 5.

Meteorite craters become invisible if their shape is changed by extensive deformation, or if they are buried by lava. The second process seems to be more important on Venus, perhaps because deformed craters are hard to recognize. Figure 8 shows an example of a crater that has almost disappeared beneath lava flows, whereas only the deepest part of the crater is Fig. 5 is flooded.

Impact craters are important because at present they provide the only method of estimating the mean age of the surface of Venus. The principle of the method used to do so is straightforward. Because of continual bombardment, the density of craters larger than some radius increases with the age of the surface. If the impact rate is known, the surface age can be estimated. Various such studies have been carried out, and show that the average surface age of Venus is about 500 million years (Ma). Surprisingly, all regions that have been dated in this way appear to be of similar age. In this respect Venus is quite unlike the Earth, whose ocean floor has a mean age of about 60 Ma, with a systematic variation from an age of 0 Ma on the spreading ridges, to about 180 Ma for the oldest ocean floor. In contrast the mean age of the continents is about 1800 Ma. If similar age variations were present on Venus they would have been discovered by counting craters. So it seems as if the entire planet was resurfaced 500 Ma ago, and that the process involved then stopped. No one yet understands how this took place, and why the history of Venus has been so different from that of the Earth.

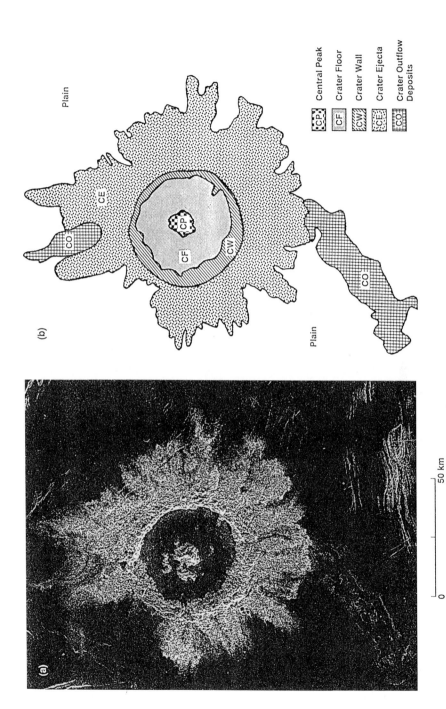

Fig. 5 Image and sketch map of the crater Danilova. Illumination is from the left with a 35° incidence angle (Weitz 1993, *in* Ford *et al.* 1993).

0   100 km

Fig. 6 Mead crater, the largest on Venus. The stripes are artefacts of the processing. Illumination is from the left with a 35°
incidence angle (Weitz 1993, *in* Ford *et al.* 1993).

Volcanism

Many parts of Venus are covered with huge lava flows. Figures 9 and 10
show images of two such features. The bright and dark regions correspond to variations in surface roughness, which varies with the speed
with which the lava flowed. So the boundaries of the bright regions do

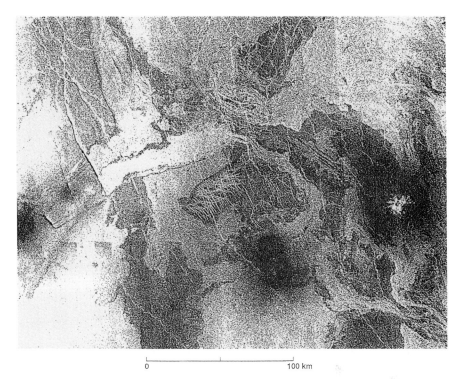

0   100 km

Fig. 7 Dark splotches formed by the breakup of meteorites in the atmosphere (Weitz 1993, *in* Ford *et al.* 1993).

not always correspond to the edges of individual lava flows. Small features such as channels, levees, and ponds of lavas caused by pre-existing features, are clear in Figs 9 and 10, because of the high resolution of the SAR images. Similar features are found on Earth. Although the flows in Figs 9 and 10 are larger than any recent flows on Earth, in the past many flows of this size or larger have been erupted. Because they have now been eroded or covered with sediment, they are less obvious on Earth than they are on Venus, where erosion and sedimentation are almost absent.

Although Magellan had no means of measuring the composition of the surface of Venus, several Soviet landers had done so earlier. They found that it consists of basalt, which is also by far the most common material that makes magma flows on Earth. It is therefore likely that the flows in Figs 9 and 10 are basaltic. The surface details of the flows are preserved so well because the surface temperature of Venus is about 430°C, so no liquid water is present to erode the surface or deposit sediment. The only sediment present is wind blown dust and dust from the breakup of

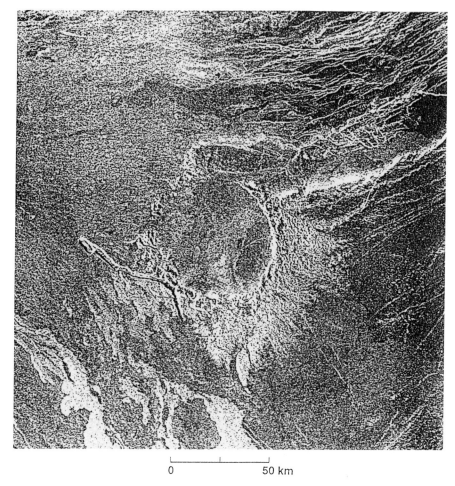

0   50 km

Fig. 8 Alcott crater, which has been almost completely
flooded with lava (Weitz 1993, *in* Ford *et al.* 1993).

meteorites. A dark patch of dust from a meteorite is visible in Fig. 9,
just outside the bottom left corner of the rectangle labelled 5.

The gradient at which a lava flow comes to rest depends on its size.
Very large flows can flow down a gradient of 1° or less, and so form
huge shallow domes. But some smaller volcanoes on Venus have produced smaller flows that form large volcanoes with steeper sides.
Figure 11 shows one of the largest, called Sapas Mons. The individual
flows are clearly visible in the SAR image. The highest parts of the
volcano rise more than 3 km above the average planetary radius. One
reason why the higher parts of the volcano in Fig. 11 are so bright is

$\overline{\text{0}}\hspace{4cm}\text{200 km}$

Fig. 9 Mylitta Fluctus, a lava delta formed from many over-lapping lava flows (Ford *et al.* 1993).

that they are rough. But all high regions of Venus are good reflectors of radar. Above a certain elevation the electrical properties of the surface change, although no one yet knows why. Its effect is to make all high mountains in the SAR images bright white, as if they were covered

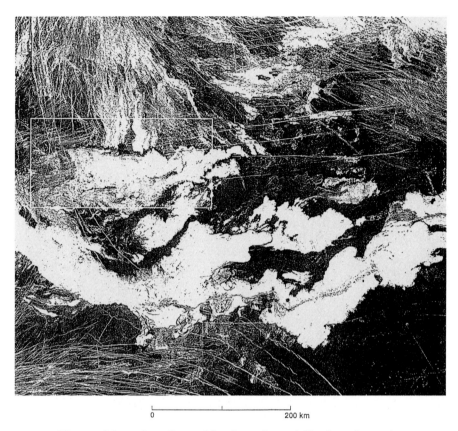

0   200 km

Fig. 10 A large lava flow with a lava channel (Ford *et al.* 1993).

in snow (which of course they are not: the surface temperature is about 430°C).

Most of the lava flows on Venus are likely to be made from basalt. However, some must have had a much higher viscosity than most basalt flows on Earth, because the edges of the flows are cliffs 100 m or more high. On Earth the viscosity of lavas is largely controlled by the SiO_2 concentration. A lava with a SiO_2 concentration of 65–75% can produce features like those in Fig. 12.

Figure 13(a) shows examples of a different type of highly viscous lava flow that is not found on Earth, called a pancake dome. These domes are formed by single eruptions, unlike the volcano in Fig. 12. The shape of these domes is that expected for a drop of very viscous material spreading over a rigid surface. Figure 13(b) compares the profile across a dome like those in Fig. 13(a), measured by the altimeter on Magellan, with the shape expected for a spreading viscous drop. The agreement between

0   200 km

Fig. 11 Sapas Mons, a large shield volcano (Ford *et al.* 1993).

the two is good, and allows the viscosity of the lava to be estimated to be about 10^{16} Pa s.

Basaltic melts sometimes intrude into planar cracks that form by brittle failure in solid rock. If the sheets are vertical they are called dykes, whereas if they are horizontal they are known as sills. On Earth basaltic dykes are 30–50 m thick, and extend for as much as 2000 km horizontally. Similar features are visible in Fig. 14(a), because a shallow linear depression forms at the surface where the two sides of the dyke separate. Figure 14(b) shows a larger-scale image of one of these depression, just south of the domes at the centre left (which are the same as those in Fig. 13a). The melt in these dykes moves at about 5 m/s when

0   100 km

Fig. 12 A large volcano constructed from very viscous lava
(Ford *et al.* 1993).

it is flowing, and takes a few days to travel 1000 km. In this time the
temperature of the melt falls by about 50°C.

Another volcanic landform that is common on Venus is formed by
flowing lava. Figure 15 shows a number of short channels that were
formed when melt flowed out of calderas. Much longer channels also
occur. Figure 16 shows a segment of the longest, whose length is almost
8000 km. Although its meanders look like those of a large river, it was
formed by flowing basalt, not water. Surprisingly, the gradient along this
channel does not slope consistently in either direction, and it is not
known which way the melt flowed.

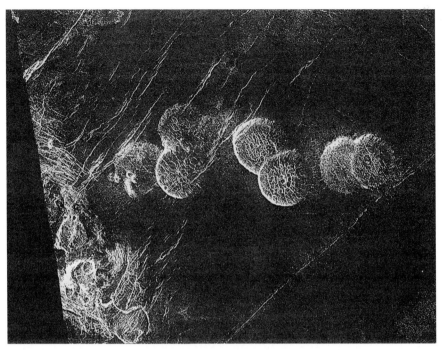

(a)

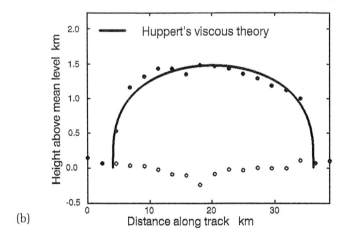

(b)

Fig. 13 (a) Overlapping pancake domes, constructed from very viscous lava. (b) The solid dots show a profile across a dome like one of those in (a), and the solid line shows the expected shape of a viscous spreading drop, calculated from Huppert's (1982) expression.

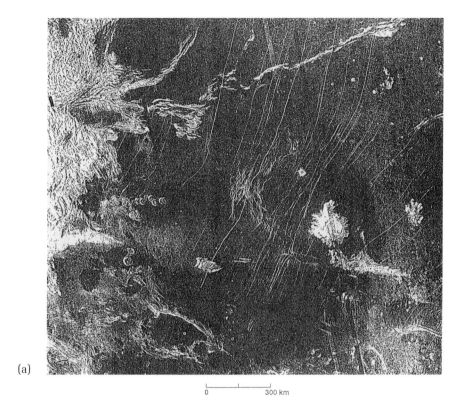

(a)

0   300 km

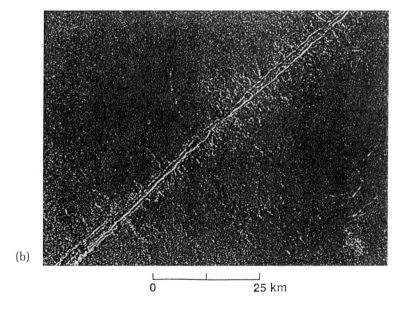

(b)

0   25 km

Fig. 14 (a) The bright linear features are the surface expression of dykes that extend for more than 1000 km across the dark plains. The domes in Fig. 13(a) are visible at the centre-left (Weitz 1993, *in* Ford *et al.* 1993). (b) a detail of one of the small grabens formed above a dyke (Stofan *et al.* 1993, *in* Ford *et al.* 1993).

0   100 km

Fig. 15 Lava channels flowing out of calderas (Ford *et al.* 1993).

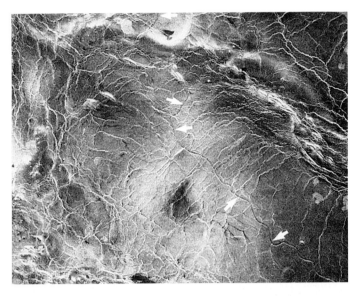

Fig. 16 A detail of the longest channel on Venus, whose course is marked by white arrows.

Tectonics

The evolution of oceans and continents on Earth largely results from plate motions. Before Magellan, various authors had attempted to identify features on Venus corresponding to the three types of plate boundary known on Earth. The Magellan SAR images showed that none of these suggestions were correct, and that features such as terrestrial plate boundaries are not common, although they do exist. The most obvious tectonic features on Venus are some large rifts that are formed by extension of the outer layers of the planet. Figure 17 shows one example. The long dark and bright linear features are faults. Where the movement produces a fault that slopes to the left, the SAR illumination produces a bright line, and the image of the fault scarp is often overlain (see Fig. 3). Fault scarps that slope to the right are not illuminated by the SAR, and show up as dark lines. These patterns can be used to estimate the amount of extension, which is only about 45 km. The rift in Fig. 17 is

Fig. 17 A large graben system produced by extension. Left facing faults show as bright lines because of overlay, whereas right facing faults are not illuminated, and are therefore dark (Stofan *et al.* 1993, *in* Ford *et al.* 1993).

0   200 km

therefore more like the East African Rift than the oceanic rifts that produce major ocean basins, where the amount of extension is often thousands of kilometres.

On Earth there are three types of plate boundaries: where plates separate from, slide past, or move towards, each other. On Venus sliding boundaries have been identified in only one small part of the planet. Structures like those formed on Earth where plates move towards each other are much commoner. It is not known whether any of these boundaries are still active, or how much movement occurred between the plates on either side. However, the distribution of craters suggests that the surface of Venus is not now being replaced by spreading ridges on a large scale. This conclusion is consistent with the rarity of features like spreading ridges in the SAR images.

Mantle convection

On Earth the plate motions themselves transport about 80% of the heat that is lost from the mantle. New plate is formed on ridges, where hot mantle upwells to fill the space between the separating plates. As the plates move sideways they cool and contract. It is this process which causes the places where plates are separating to form ridges, and the depth of the ocean to increase with age. Where plates are destroyed, by one plate overiding the other, one sinks into the mantle. Because it is colder than its surroundings, it is denser and hence sinks under its own weight. This process of cooling at the surface, followed by the creation of cold sinking regions by plate destruction, is called thermal convection, and is familiar to all cooks! It can occur in the mantle, even though the mantle is made from solid rock, because solids at temperatures close to their melting point can flow. Such flow is called creep. The best known example of such flow is a glacier, which moves downhill even though it consists of solid ice, because the ice is close to its melting temperature and can therefore creep. The rock of the mantle also flows because its temperature is close to its melting temperature.

On Earth the mantle circulation is dominated by plate creation and destruction. There are, however, places like Hawaii where a different type of flow occurs that is largely unaffected by the plate motions, and where a hot plume rises beneath the interior of a plate. Where such plumes approach the surface, their temperature exceeds the melting temperature and large volumes of melt are produced. It is this melt that has constructed the islands of Hawaii. Figure 18 shows a section through a convective model of Hawaii that matches the observations. As

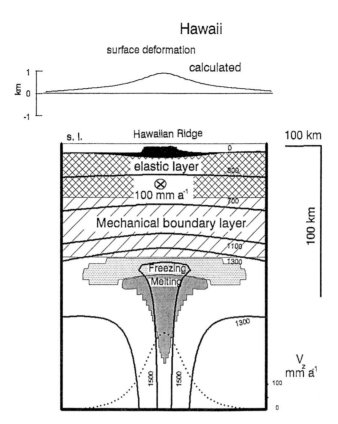

Fig. 18 A model of the axisymmetric plume beneath Hawaii. The numbers show the temperature of the isotherms in °C. The vertical velocity V_z at the base of the figure is shown by the dotted line.

well as generating the melt that forms the Hawaiian Ridge, the hot rising plume pushes the plate upwards to form a long wavelength bulge in the sea floor. Although features like Hawaii are impressive, the associated plumes only transport about 10% of the heat lost from the mantle, or much less than do plate motions. None the less such features are important, because they show how the mantle circulates when it is not being driven by plate motions. An important means of mapping such flow uses the gravity field, because hot rising plumes produce small long wavelength positive gravity anomalies. Figure 19 shows how they do so. The plume itself is hot, so its density is less than that of the surrounding mantle because of thermal expansion. It therefore produces a negative gravity anomaly. But it pushes up the plate above it, and the extra material produces a positive gravity anomaly, which is slightly larger than the

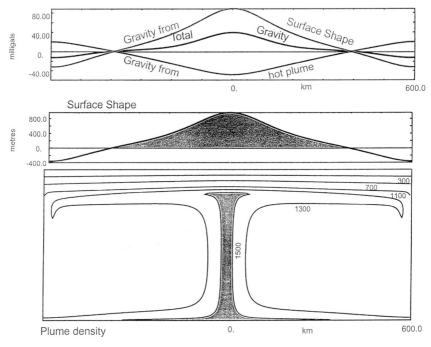

Fig. 19 The total gravity anomaly above a convecting plume results from a positive anomaly from the surface deformation, which is larger than the negative anomaly from the hot plume.

negative anomaly from the plume. The total gravity anomaly is therefore positive, and about a factor of three smaller than would be expected from the topography alone. On Earth variations in crustal thickness dominate the surface topography, which therefore cannot be used directly to map the convective circulation. However, crustal thickness variations are compensated. Continents have a crustal thickness of about 33 km, compared with 7 km for oceans and up to 70 km for high mountains. When the topography is compensated by variations in the thickness of the low density crust, the resulting gravity anomalies are small; much smaller than those due to the plumes. So the gravity field can be used to map the geometry of mantle convection and to distinguish those topographic features that are maintained by convection from those that are supported by crustal thickness variations.

The same approach can be used on Venus if the gravity field can be mapped. Fortunately the gravity field affects the orbit of any satellite, like Magellan, that is in orbit round Venus. Because Venus is so far from the Earth, the orbit of Magellan cannot be obtained from the pointing direction of the big dishes of the DSN. Instead the Doppler shift of the

carrier frequency is used to measure the line of sight velocity of the
spacecraft, to an accuracy of about 0.1 mm/s. This velocity is then used
to model both the Magellan's orbit and the gravity field of Venus. Figure
20 shows maps of the gravity field and of the shape of the planet. The
third map shows what remains when the convective topography corresponding to the gravity field is subtracted from the observed topography. The difference is called the residual topography. Regions with
positive residual topography are like continents on Earth: because they
lack gravity anomalies, the surface elevation must be maintained by an
increased crustal thickness, rather than by mantle convection. The large
positive gravity anomalies in Fig. 20(a) are all associated with elevated
topography, and show the locations of rising plumes. They are surrounded by linear negative anomalies, which mark regions where colder
mantle is sinking. Therefore, the gravity field from Magellan directly
maps the mantle circulation within Venus. Because of the absence of
plate motions, and because the viscosity of the convecting region is
about 10 times greater than it is on Earth, the map in Fig. 20(a) is easier
to interpret than are corresponding maps for the Earth.

The large size of many of the convectively supported topographic features on Venus compared with those on the Earth is surprising. It suggests a possible explanation for the puzzling gradients in the long
channel in Fig. 16. Convection in the mantles of both Earth and Venus is
likely to be time dependent, partly because new plumes start from the
bottom of the convecting region, and partly because the whole convection pattern slowly changes. the surface topography supported by convection must therefore also change with time, and the gradients along
the channel may also change after the channel itself has formed. This
idea is easily tested, by generating a profile of the residual topography
along the channel. Figure 21 illustrates profiles of the measured and
residual topography along the channel, and shows that the long wavelength reversals of gradient are indeed the result of convection. The
shorter wavelength irregularities correspond to features on the SAR
images that are produced by thrusting.

The gravity field can also be used to measure the thickness of the
elastic layer that supports short wavelength topography on both Earth
and Venus. Although the method used is complicated, its principle is
simple. If a volcano is constructed from basalt, whose density is less
than that of the mantle, on top of a thick rigid plate, it produces a large
positive gravity anomaly, because of the extra mass of the volcano. If,
however, the elastic part of the plate is thin and bends under the weight
of the volcano, the topography is compensated by the low density root of
basalt, and the gravity anomaly is much smaller. Therefore in principle

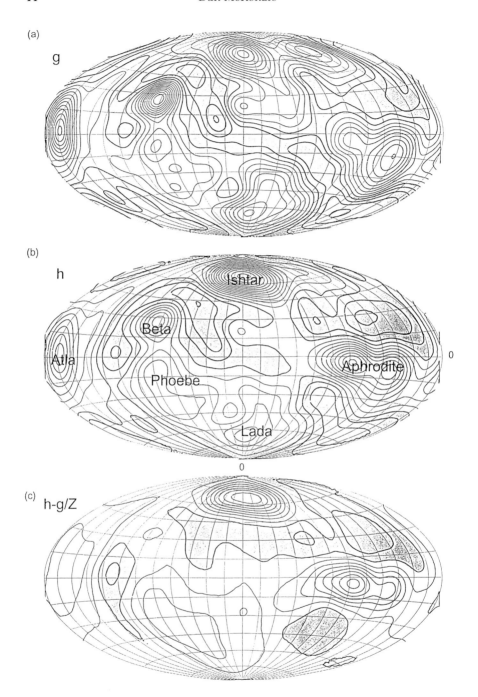

Fig. 20 (a) Gravity, (b) topography and (c) that part of the topography which is not convectively supported (known as the
residual topography).

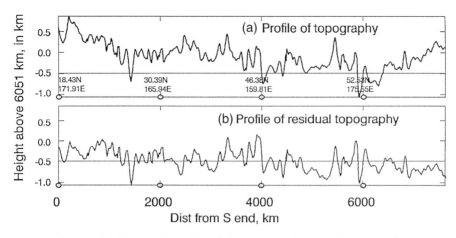

Fig. 21 (a) Observed profile of the topography, and (b) residual topography, along the channel shown in Fig. 16.

the relationship between gravity and topography can be used to estimate the thickness of the outer elastic layer on both Earth and Venus. It is not straightforward to use this method on Venus, because the short wavelength gravity anomalies of interest are not measured accurately by Magellan. Complicated processing of the data is required to extract the signal from the noise. When this is done the elastic thickness is found to be about 30 km, or very similar to that for Hawaii, even though the surface layers are hotter than they are on Earth.

Figure 22 shows a section through a plume on Venus for comparison with that shown in Fig. 18 for Hawaii. There are a number of important differences between plumes on Earth and Venus. On Venus the surface layer that forms the lid on top of the convecting region is thicker than it is on Earth, even though the surface temperature is 430°C instead of °C. Also, the Venusian plume is larger than that beneath Hawaii, because of its higher viscosity.

Venus and Earth

Magellan has given us our first look at a planet that is similar to the Earth, and whose mantle is still actively convecting. As we expected before the spacecraft was launched, the gravity field has given us most information about what is going on now, and we have (just!) enough information to construct a convective model of the mantle circulation of Venus at the present day. But as yet we understand almost nothing

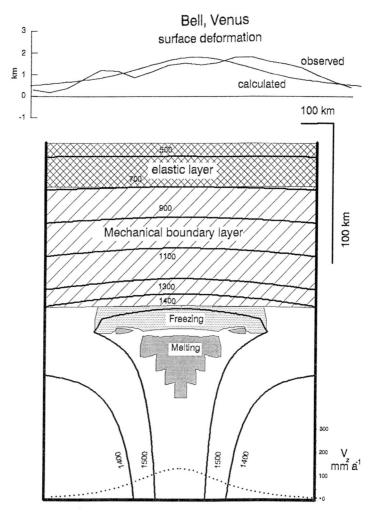

Fig. 22 A model of a venusian plume (compare with Fig. 18). The numbers show the temperature of the isotherms in °C. The vertical velocity V_z at the base of the figure is shown by the dotted line.

about its geological history. The major difference between the Earth and Venus is that the mantle of Venus is more resistant to deformation than is that of the Earth at the same temperature. This difference allows Venus to support elastic stresses at temperatures below about 650°C, compared with an upper limit of about 500°C for the Earth, and causes the viscosity of the mantle to be about a factor of ten greater. The same effect can also account for the absence of large scale plate tectonics on

Venus, by increasing the strength of faults. The most probable cause of this difference is that Venus contains much less water than does the Earth, both in its atmosphere and mantle. Even though the water content of the Earth's mantle is only about 50–100 parts per million, its presence has a major influence on the strength and viscosity of the mantle. It is not yet clear whether the water in the Earth's mantle has remained there since the Earth formed, or whether it has been transported downwards from the oceans by sinking plates. Nor is it clear why Venus is now so much drier than the Earth. None the less the fact that Venus *is* dry can account for the principal differences between it and the Earth.

When I explained all this to a friend of mine who is a geochemist he said 'so all you have discovered by spending $1000 million is that Venus is dry.' In one sense his implied criticism is fair. However, until Magellan no-one had any idea that the small amount of water in the Earth's mantle controlled its tectonics. We have only understood this because of the Magellan mission. Whether the mission was 'worth $1000 million' is a question that I am often asked. It is one which I can see no sensible way to answer. I am glad that such decisions are made by politicians, and not by me. But what is clearly silly is to spend so much money on the mission and so little on analysing the data. I have been able to obtain about £10 000 a year to do so with Francis Nimmo, a graduate student at Cambridge with whom most of this work was done. I am grateful to the Royal Society and NERC for this support. Even such modest amounts of money have not been available to most of the US planetary geologists working on Venus.

Notes and bibliography

Most of the data from the Magellan project is discussed in two special issues of the *Journal of Geophysical Research (Planets)*, **97** (1992), 13063–13689 and 15921–16380. Many of the papers in these issues are rather technical.

Ford J.P., Blom R.G., Crisp J.A., Elachi C., Farr T.G., Saunders R.S., Theilig E.E., Wall S.D., and Yewell, S.B. (1989). *Spaceborne radar observations: a guide for Magellan radar-image analysis*, JPL Publication 89–41 (12/89), Pasadena, CA explains how to interpret SAR images, and how various artefacts arise. It was written before any Magellan images were available.

Most of the images in this chapter are taken from Ford J.P., Plaut J.J, Weitz C.M., Farr T.G., Senske D.A., Stofan E.R., Michaels G., and Parker T.J. (1993). *Guide to Magellan image interpretation*, JPL Publication 93–24, Pasadena, CA. The images themselves are available on CD-ROMs from NSSDC (see the web page http://nssdc.gsfc.nasa.gov/cd-rom/cd-rom.html NSSDC CD-ROM CATALOG Dec 1996).

Roth E.L. and Wall S.D. (ed.) (1995). *The face of Venus*, NASA SP-520 Washington DC, is a photo-atlas with many spectacular images of Venus.

More technical accounts can be found in Venus II, shortly to be published the University of Arizona Press, which reviews much of the work that has been carried out on the Magellan data. A shorter review, by Nimmo F. and McKenzie D. (1998) Volcanism and tectonics on venus, *Annual Reviews of Earth and Planetary Science*, **26**, 23–51, is especially concerned with the gravity observations.

DAN MCKENZIE

Born 1942, graduated in Physics from the University of Cambridge, and completed a PhD at Cambridge under Sir Edward Bullard. He then worked at a number of oceanographic laboratories in the US before returning to Cambridge. He is now a Royal Society Research Professor and a Fellow of King's College. His present interests are planetary geology and melt generation, but is best known as one of the originators of the theory of plate tectonics. He was elected to the Royal Society in 1976, and also belongs to the US National Academy of Sciences.

Meteorites: messengers from the past

MONICA M. GRADY

Abstract

Meteorites are an important part of our Solar System. They have a direct effect on the Earth. Our planet is bombarded constantly from space by the tiniest of dust grains, and less frequently by enormous impact crater-forming bodies. Meteorites are ancient, older than the oldest rocks now present on Earth, and are credited with bringing both life and death to our planet. It is possible that organic molecules and water were introduced to the early Earth by meteorites and comets, sowing the seeds which eventually led to the evolution of life in its myriad of forms. But meteorites also have an awesome potential for destruction: a huge body fell at the end of the Cretaceous (65 million years ago), resulting in a large crater, and inadvertently led to the demise of the dinosaurs and many other groups. Meteorites fall all over the Earth, and have been collected from the icy wastes of Antarctica, the hot sands of the Sahara, and even the back garden of a senior citizen in middle England. The location and timing of meteorite falls are unpredictable: we do not know where or when the next one will land.

This chapter explores the nature of meteorites, from their formation at the birth of the Solar System to their final resting place on Earth. It highlights some of the different types of meteorites: the most primitive stony meteorites, which are related to comets and contain 'stardust' produced in the out-flowing wind of ancient stars; the dense iron meteorites, some of which are the nearest accessible analogues to the Earth's core, and the group of igneous meteorites thought to come from our neighbour, Mars, and have helped us to unlock the secrets of the Martian surface.

Introduction

Meteorites are fragments of ancient material, natural objects that survive their fall to Earth from space, and are recovered. They are not radio-active, and are almost always cold when they land. Meteorites are distinguished from *meteors*, or 'shooting stars', which are pieces of dust that burn up high in the atmosphere. No material is recovered on Earth from a shooting star. A meteoroid is a small body travelling through space, that may, or may not, land on Earth as a meteorite. All these words derive from the Greek '$\mu\varepsilon\tau\varepsilon\omega\rho\sigma\varsigma$', meaning 'things on high', the same root as meteorology, the study of the weather (a truly atmospheric phenomenon). The study of meteorites is known as *meteoritics*.

Meteorites were formed at the birth of the Solar System, ~ 4560 million years ago. Although the Earth, along with the other planets, was also formed at this time, none of the original material remains: it has been removed by bombardment or otherwise eroded, or recycled through geological activity (plate tectonics, volcanism, etc.). It is only by studying meteorites that we can learn about the processes and materials that shaped the Solar System and our own planet.

Meteorites are divided into three types on the basis of their composition: stones (composed of minerals often found in rocks on Earth); irons (made up of iron metal alloyed with nickel); and stony-irons (as their name suggests, a mixture of stone minerals and iron metal). A distinction is made between meteorites which are observed to fall (mostly stony meteorites, reflecting their abundance on meteorite parent bodies) and those that are found. The latter tend to be mainly iron meteorites, because they readily stand out from surrounding terrestrial rocks. Meteorites are generally named after a place near to which they fall or are found (usually the nearest Post Office). Exceptions occur in desert regions, where locality names are few; in these cases, meteorites are given the name of the geographical area in which they were found, followed by a number.

How big is a meteorite and how often do they fall?

Meteorite falls are not rare events: in fact, they are a lot more common than people realize. Approximately 40 000 tonnes of extraterrestrial material fall on the Earth each year—this is about four particles per hour per square km of the Earth's surface. Fortunately, almost all of this material arrives as dust, (known as micrometeorites or cosmic dust), small fragments (< 1 mm across) that are captured by the Earth and come from asteroids and comets. A small proportion of micrometeorites might also be from outside the Solar System, from interstellar space. The very smallest of micrometeorites (< 50 μm) do not melt as they pass through the Earth's atmosphere, but remain as fluffy aggregates of silicate minerals. This cosmic dust has been collected by aircraft equipped with specialized collector plates, flying high in the stratosphere. Slightly larger particles, like the one in Fig. 1, melt as they fall and form tiny rounded droplets. Micrometeorites such as these 'cosmic spherules' were first separated from deep sea sediments collected from the ocean floor by the *HMS Challenger* expedition of 1872–76. Subsequently, cosmic spherules and unmelted micrometeorites have been collected by filtering the water produced from melting large volumes of Antarctic ice, and also from melt-water streams in Greenland.

Generally speaking, larger meteorites fall less frequently than small ones. Meteor Crater in Arizona (Fig. 2) was produced by the impact of

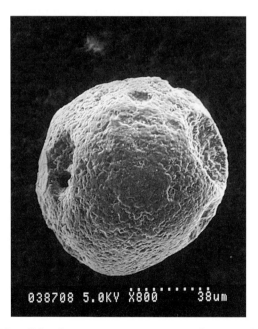

Fig. 1 The slide shows a micrometeorite photographed in a scanning electron microscope. The grain is ~ 100 μm in diameter, and has been magnified 800 times. The micrometeorite was one of several hundred recovered by melting a large volume of Antarctic ice, and filtering the resulting water. (Photograph Courtesy of Dr M.J. Genge, Natural History Museum.)

Monica M. Grady

Fig. 2 Meteor crater in Arizona, ~ 1.2 km across and ~ 200 m deep. It was formed by the impact of an iron meteorite, ~ 5–40 m across, approximately 50 000 years ago. (Photograph courtesy of USGS.)

an iron meteorite ~ 50 000 years ago. The crater is ~ 1.2 km across; the original meteorite was estimated to weigh between 15 000 and 25 000 tonnes, and would have been ~ 35–40 m in diameter, but most of it was vaporized by the impact. Over 30 tonnes of iron meteorite, named *Cañon Diablo*, have been recovered from the vicinity of Meteor Crater. Any environmental effects associated with an impact this size would be mostly localized or short-term. For instance, living creatures in the vicinity of the impact site would have been killed instantly, but the impact would not have led to the world-wide extinction of any species. A more extensive phenomenon associated with an impact of this magnitude might have been reddening of sunsets due to dust in the atmosphere, carried very quickly around the globe, but this would rain out on a time-scale of a few weeks or months, before it could have a significant effect on global temperature. Impacts of this size occur approximately every 10 000 years. Larger impacts occur even less frequently: every 50 million years or so. A huge bolide fell at Chicxulub, in the Yucatan Peninsula in the Gulf of Mexico at the end of the Cretaceous period (65 million years ago). The crater is now buried, but geophysical surveys estimate its diameter to be between 180 and 320 km. Environmental effects caused by an impact of these dimensions include a darkening of the sky, due to ejected rock dust, followed by a rapid, global drop in temperature. In the case of Chicxulub, the impact was into a sedimentary rock formation, including evaporite deposits (i.e. sulphate-bearing rocks), resulting in tonnes of sulphur oxides ejected into the atmosphere. The energy of the impact fused nitrogen and oxygen from the atmosphere into nitrogen oxides. As the temperature dropped, sulphur and nitrogen oxides washed out of the atmosphere as acid rain, leading to a change in the pH (acidity) of the oceans. These consequences affected the entire globe, not just the local region, and for an extended period of time. It is entirely possible that the end result of these global environmental changes was the extinction of many species, including the dinosaurs, although this is by no means accepted by many palaeontologists.

Between the micrometeorites and the crater-forming meteorites lies the range of 'recoverable' meteorites. Several thousand meteorites, about the size of a football, fall on the Earth every year. Many break up in the atmosphere, and three-quarters fall in the oceans and are lost. Usually, only five or six are seen to fall and end up in collections. No one is known to have been killed by a falling meteorite, apart from the alleged death of a dog by a stone from the *Nakhla* Martian meteorite that fell in Egypt in 1911, although several people have been injured. For example, a young boy was hit on the arm by a meteorite while playing football in the town of Mbale in Uganda in 1992. He was fortunate to suffer only a bruise from the 30 g stone that caught him after it ricocheted off the leaf of a banana tree: the *Mbale* shower consisted of several tens of stones, the largest of which weighed 27 kg. If the boy had been hit by any of the larger stones, he would probably have been killed. Meteorites can, however, do much damage to property: several have hit houses and broken roofs and windows, and cars are also regular targets, for example, the *Peekskill* meteorite fell in New York State in October 1992, landing in the boot of a parked car.

The last meteorite to fall on England landed in 1991, in Glatton, near Peterborough, in the garden of a Mr Pettifor. It is a stone meteorite, and weighs almost 0.7 kg. It fell through a hedge of conifers, and landed below a wooden fence. It was recovered immediately after it landed, but only because Mr Pettifor was in the garden at the time. If he had been inside his house, he would not have heard the whining noise made by the meteorite as it fell through the air. This startled him, and caused him to investigate the commotion made by the meteorite as it subsequently crashed through the hedge. The meteorite is now part of the UK national meteorite collection, curated at the Natural History Museum in London.

Where are meteorites found?

Meteorites fall almost randomly over the Earth's surface, but often are lost—by falling into the ocean, or among other rocks. It is not possible to predict where and when the next one will land. Many meteorites are recovered from deserts (both cold and hot), as the dry environment ensures their preservation, and the lack of vegetation and other rocks enhances their chance of being found.

More meteorites have been found in Antarctica than anywhere else in the world. This is not because more meteorites fall in Antarctica: there is, in fact, a slight decrease in the flux of meteorites at the poles compared with the equator, but because those that do fall are preserved in the ice, for as long as a million years. As the ice moves towards the edges of the continent, the meteorites are carried along. When the ice meets a barrier that it cannot cross (such as a mountain chain), the ice is forced upwards. Subsequent ablation (or stripping) of the ice is caused by the winds that constantly blow downwards from the South Pole, scouring the ice surface. An equilibrium is reached, between flow of ice outwards and removal of ice by ablation. The continual removal of ice results in a gradual build-up of the solid materials it carries with it (mostly meteorites, but also including fragments of scoured bedrock), leading to concentrations of meteorites in these so-called 'blue ice' areas (Fig. 3). The first concentration of meteorites in Antarctica was recorded in 1969. Since then, meteorite collecting expeditions by US, Japanese and European scientists have returned ~ 15 000 specimens, representing approximately 8000 individual meteorites. Included in this total are some of the rarest of all types. The total compares with ~ 2000 'regular' meteorites from more temperate latitudes amassed in the 200 years since meteorites were recognized as natural phenomena. The success of the Antarctic collection has necessitated a change in the procedure for naming meteorites: Antarctic meteorites are named by giving them a number detailing the year of the field expedition that recovered them, followed by a curatorial analysis number. The whole is prefixed by an abbreviation of the field area, e.g. ALH 84001 was recovered in 1984 from the Allan Hills region of Antarctica, and was the first specimen examined on its return from the field.

Meteorites have also been found in several hot deserts, for example the Sahara in Africa and the Nullarbor Plain in Australia, in all cases on hard, rocky surfaces, rather than sand. Although they are not concentrated by a specific mechanism, such as the ice movement described in Antarctica, many meteorites have been recovered from hot deserts, as these deserts are relatively more accessible than the Antarctic. Meteorites

Fig. 3 A large stony meteorite found on the ice in the Lewis Cliffs area of Antarctica ($\sim$ 200 km north of the South Pole, near the Beardmore Glacier in the Queen Alexandra mountain chain). The dark surface of the meteorite is readily visible against the blue background of the ice. The meteorite weighs about 11 kg.

that have fallen over the past 50 000 years are preserved on these old land surfaces and are readily distinguished from the surrounding terrain. Again, they are given a combined geographic and numeric name.

Where do meteorites come from?

Our local star, the Sun was formed out of a nebula, a rotating cloud of gas and dust. As the cloud rotated faster, it collapsed and flattened into a disk with the Sun at its centre. Within the disk, dust grains joined together to form bigger and bigger bodies, eventually producing the planets. The Sun and planets formed approximately 4560 million years ago. There are many active regions throughout the universe where star formation is still occurring, e.g. the Orion Nebula (Fig. 4).

Fig. 4 The Orion Nebula (750 000 million km or 1500 light years away) is visible (through binoculars) as a glowing cloud just below the 'belt' of the constellation of Orion. The area imaged is 6500 million km across, and shows an active region of newly forming stars. It is likely that our Sun, like many stars, formed in an environment similar to that of the Orion Nebula. (Photograph courtesy of Space Telescope Science Institute.)

The Sun is the star at the centre of the Solar System, and all the planets orbit around it. The four inner planets (Mercury, Venus, Earth, and Mars) are mainly made from rock. Then follow the giant planets Jupiter and Saturn (composed mainly of gas) and then the outer planets Uranus and Neptune (gas and ice). The outermost planet, Pluto, and its satellite Charon, are small compact icy bodies that have an affinity with a disk-like array of similar objects known as the Kuiper Belt. There are thought to be approximately 35 000 Kuiper Belt objects orbiting the Sun at the distance of Pluto (5913 million km or ~ 40 AU; 1 AU is the mean Earth–Sun distance, approximately 150 million km) and beyond. When objects from this belt are dislodged by gravitational perturbations, they might enter the inner Solar System as comets. Between Mars and Jupiter, at the hiatus between rock and gas in the Solar System, lies the Asteroid Belt, the place from which most meteorites come.

The asteroids lie in the region between Mars and Jupiter and orbit the Sun at a distance of approximately 450 million km, which is three times that of the Earth from the Sun (3 AU). There are several thousand aster-oids, the largest of which, named Ceres, is ~ 914 km across (for compar-ison, Earth has a diameter ~ 13 000 km, and Moon ~ 3500 km). Asteroids are rocky, metallic or carbonaceous bodies. They are material remaining after the planets formed: Jupiter's gravitational pull prevented the bodies from joining together to form a single planet. Occasionally, influenced by Jupiter, the orbit of an asteroid is altered such that it might collide with another, and break up. Images of asteroids obtained recently by the Galileo probe show that asteroids themselves have cratered surfaces indicating that collisions are frequent within the Asteroid Belt. Fragments of disrupted asteroids fall to Earth as meteorites.

How is it known that most meteorites originate from the Asteroid Belt? One way is to photograph incoming meteoroids and measure their speed and entry path—the orbit can then be calculated. There have been three separate investigations of this type, each involving a network of ground-based cameras. For example, the Canadian Meteorite Observation and Recovery Program operated from 1971 to 1985, photographing the night sky using an automatic camera and taking photographs at 0.25 second intervals. As a result of this programme, the fireball of the *Innisfree* meteoroid was photographed in 1977 (Fig. 5), and 3.79 kg of meteorite recovered as six specimens and several fragments. The main piece was found 12 days after the fireball observation, on snow.

One of the most recent meteorite falls to be recorded was that of *Peekskill*, to which reference was made in the second section, as it landed in the boot of a car. The track of the extremely bright fireball was recorded on several video cameras, mostly by members of the public attending outdoor football games on the Friday afternoon of the event. The video footage has been edited together to produce a film of the fireball travelling over the north-eastern USA. It is clear that the mete-oroid broke into several pieces during its flight, each of which became a fireball. Only one meteorite was recovered, and that was the 12.4 kg object that hit the car. Orbits of the *Innisfree* and *Peekskill* meteoroids have been calculated, as have those of two other observed fireballs asso-ciated with the *Pribram* (Czech Republic) and *Lost City* (USA) mete-orites, and all four orbits are elliptical and extend into the Asteroid Belt, confirming that this is from where the meteorites originated.

In addition to meteorites from the Asteroid Belt, there are currently 18 meteorites from the Moon in the world's collections. Lunar meteorites can be compared directly with samples brought to Earth by the Apollo and Luna missions between 1969 and 1976. The surface of the Moon is covered in craters caused by impacting bodies. If the impactor arrives with the requisite velocity and on a favourable trajectory, then the force

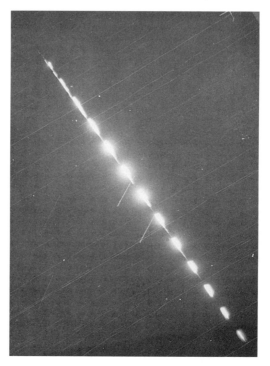

Fig. 5 A fireball photographed by the Canadian Meteorite
Observation and Recovery Program, using an automatic camera.
The photographs were taken at 0.25 second intervals. The
Innisfree (Alberta) meteorite was recovered as a result of this
observation. (Photograph courtesy of CMORP.)

of the impacts will be sufficient to eject material from the surface with a
velocity great enough to overcome the Moon's gravity and be launched
into space. Subsequently, the material goes into orbit in interplanetary
space, and some of it eventually lands on the Earth as a meteorite. In the
same way, rocks have come to us from Mars: we have 12 meteorites that
have been ejected by impact from the surface of our neighbouring planet
(see later—What can we learn from meteorites?).

Comets are also part of our Solar System. They formed at the cold
outer reaches of the Solar System, where ices could condense. The icy
bodies that produced comets are now thought to inhabit a spherical
region of space extending to ~ 50 000 AU, called the Oort Cloud. Other
icy bodies formed in the Kuiper Belt, at or beyond the orbit of Pluto.
Like the planets, comets orbit the Sun: for example, Halley's comet takes
76 years to complete one orbit. Comets have been described as 'dirty
snowballs', a mixture of ice and dust that has never completely melted.

Each time a comet approaches the Sun, part of its ice melts, and streams away from the central portion, or nucleus of the comet, carrying with it some of the dust. This expanding cloud of gas and dust gives rise to a comet's characteristic appearance of a head and a tail. Each time the comet draws near to the Sun, more of the ice and dust is lost. The dust becomes spread out along the entire orbit of the comet. For several comets, we can predict when the Earth will pass through this dust, giving us a brilliant display of shooting stars, or a meteor shower, e.g. the Perseid meteor shower that can be observed on clear nights between 9 and 12 August. It is also possible that comets are the parent objects of a group of five very special meteorites (see later—What can we learn from meteorites?)

The bright fireball often associated with an incoming meteoroid is the result of frictional heating as the body travels through the atmosphere. Only the outermost surface melts; the resulting droplets of molten meteorite are carried away by the speed of passage. Finally, as the meteoroid is slowed down by the atmosphere, the molten surface cools rapidly to a glassy coating, or fusion crust (Fig. 6). The presence of a fusion crust is often characteristic of meteorites. It is very important to stress that it is only the outermost surface of the meteorite that melts: the interior remains cool and unchanged. Meteorites are cold when they land.

Fig. 6 The very glossy black fusion crust on the outer surface of a broken stone of the *Stannern* achondrite. The fusion crust is less than 1 mm thick, and the unchanged, pale grey, basaltic meteorite is readily distinguishable from the glassy crust.

What are meteorites made from?

Meteorites are divided into three main types (stone, iron and stony-iron), reflecting their composition. Most meteorites are stony (96% of all falls), made up of the same minerals (olivine, pyroxene, plagioclase) as many terrestrial rocks, minerals which contain silicon, oxygen, magnesium, iron, calcium, and aluminium. The stony meteorites can be subdivided into chondrites, those which have remained unmelted since formation (or aggregation) of the parents. These almost all contain small rounded droplets of once-molten material, or chondrules (from the Greek '$\chi o\nu\delta\rho o\varsigma$', meaning 'granule' Fig. 7). Chondrites retain a chemical signature close to that of the original material from which they aggregated. The other division of stony meteorites is the achondrites. These are igneous rocks, like basalts, that formed from melts on their parent bodies. Achondrites do not contain chondrules, and, as a consequence of the melting process, are chemically differentiated, i.e. no longer exhibit a primordial signature. Figure 6 is of the Stannern achondrite, a typical basalt from the Asteroid Belt.

The other large division of meteorites, the irons are made dominantly from iron metal typically with 5–15 wt. % nickel. These meteorites have all been formed during extensive melting processes on the parent bodies

Fig. 7 A slice of rock about 30 μm thick, viewed in plane polarized light through a petrologic microscope, showing the appearance of chondrules. Field of view $\sim$ 2 mm.

from which the meteorites originated. The heat source for melting was, in some cases, the result of impacts, but for many iron meteorites the heat source was most probably from the decay of short-lived radioactive isotopes, such as ^{26}Al. The iron meteorite parent asteroids were sufficiently large that this heat built up and was retained, allowing reduction reactions (similar to smelting in a blast furnace) to occur within the parents. Iron–nickel metal, produced from the reduction of silicate minerals, migrated under gravity to the centre of the parents, forming a core, while the less dense remaining silicates rose to the surface, forming a crust. Iron meteorites are the closest physical analogy we have to the material which forms the Earth's core. (In contrast, either short-lived radioactive elements were absent, or heat from radioactive decay of these elements was dissipated in the smaller chondritic parents, thus melting did not occur.)

Figure 8 shows a slice of the *Gibeon* iron meteorite. The slice has been polished, then etched with acid to reveal a pattern, the Widmanstätten pattern, named after the scientist who first described it. This pattern is

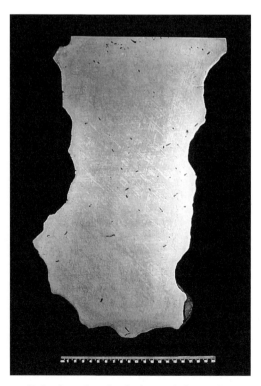

Fig. 8 A polished and etched slice of the *Gibeon* meteorite (from Namibia), showing its Widmanstatten pattern.

characteristic of most iron meteorites, and is produced from the inter-growth of two alloys of iron metal with nickel (kamacite and taenite), each alloy containing different nickel concentrations. It is a result of the very slow rate at which the metal cooled: between 1 and 100°C per million years, allowing nickel atoms to diffuse through the iron lattice. The final pattern is 'frozen' in when the nickel no longer has sufficient energy to move. The slice also shows dark areas, which are patches of iron sulphides and elemental carbon as graphite.

The final main subdivision of meteorites is the stony-irons: a mix, as the name suggests, of stone and metal. The pallasite subgroup of these very rare meteorites, composed of almost equal volumes of stone and iron, have one of the most beautiful of appearances, produced from the intergrowth of iron–nickel metal with olivine (a magnesium–iron sili-cate mineral, common on Earth as the semiprecious gemstone, peridot). Pallasites were also formed by melting in their parent, and represent an intermediate stage between iron meteorites and differentiated silicates, a snapshot of material from the core/mantle boundary of the body. Figure 9 is a slice of the *Brenham* pallasite.

Fig. 9 A polished slice of the *Brenham* pallasite (from Kansas, USA), composed of orange-brown olivine grains set in a network of iron–nickel metal.

What can we learn from meteorites?

Different types of meteorite provide evidence about events that have occurred as the Solar System formed and evolved.

One important class of stony meteorites is that of the carbonaceous chondrites. These have chemical compositions (apart from hydrogen and helium) that are close to that of the Sun. The parent bodies of the carbonaceous chondrites probably formed towards the outer edges of the Asteroid Belt, where it was cooler, allowing some ices to condense. The carbonaceous chondrites can be subdivided into several groups, the most primitive of which is that of the CI chondrites, very rich in water, sulphur, and organic compounds. There are only five of these rare meteorites, and it is possible that this small group might be from the remains of burnt-out comets, rather than from the Asteroid Belt. Each time a comet approaches the Sun, it loses ice. It is possible to envisage an episode when the last vestiges of ice are evaporated from a cometary nucleus, leaving behind just the stony material. This, then, having no 'fuel' to drive it, falls towards the Sun, and may be captured by the Earth as an unusual meteorite.

There are several other groups of chondritic meteorites, also with compositions unfractionated since their aggregation, but with lower volatile contents. Although these meteorites have not melted since their formation, they do contain materials that were once molten (Fig. 7). These spherical silicate assemblages (chondrules) were produced by rapid cooling of droplets of molten stone. The droplets came from collisions between clumps of dust grains in the early stages of the formation of the Solar System, so meteorites such as these represent the materials from which the Solar System grew.

In addition to chondrules, chondrites contain organic compounds in varying quantities. Some groups contain amino and carboxylic acids and complex hydrocarbons, while others contain elemental carbon. It is meteorites like these, together with the ice- and volatile-rich comets, which probably brought volatile materials to the newly formed Earth, and helped establish our planet's atmosphere and oceans. Without them, there would be no life on Earth.

Also buried within chondrites are tiny grains of dust that came from stars other than our own Sun. These grains are diamonds (invisible to the naked eye, only 3 nm across) and silicon carbide (or carborundum). The diamond grains in Fig. 10 are too small to be seen individually. They were separated from the meteorite by dissolving 10 g of sample in strong acids, to remove all the stony material, leaving behind clumps of an acid-resistant residue containing diamonds, weighing only ~ 40 μm

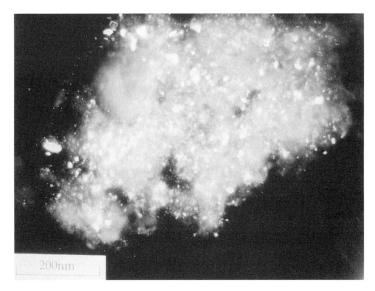

Fig. 10 A clump of interstellar diamonds isolated from a meteorite by dissolving away all of the stony and metallic components. The individual diamond crystallites are too small to be seen in this image taken using a transmission electron microscope. The material has been magnified approximately 150 000 times (Photograph courtesy of Dr M. Lee, University of Edinburgh.)

and thus representing about 4 p.p.m. of the starting material. The interstellar origins of the diamonds have been inferred from the isotopic compositions of nitrogen and the noble gas xenon trapped within the diamond lattice, and released when the diamond is burnt in the laboratory. The diamonds were blown from the surfaces of neighbouring stars, and carried on the stellar wind into the collapsing dust cloud that formed our Solar System. From these grains, we learn that our Sun did not grow in isolation, but had neighbours.

In an earlier section ('Where do meteorites come from?'), it was reported that almost all meteorites come from the Asteroid Belt, and that a few come from the Moon. There are also 12 meteorites from Mars. How is it known that they come from Mars? We know much about the composition of Mars' atmosphere and surface from data from two NASA spacecraft that visited the planet in 1976 (the *Viking* probes). Experiments on board the craft measured the atmosphere as they fell towards the planet. The probes also scratched the surface and analysed the composition of the soil. In 1979, the 8 kg stone meteorite EET A79001 (Fig. 11) was found in the Elephant Moraine region of Victoria Land in Antarctica.

Fig. 11 The EET A79001 martian meteorite (8 kg in weight), in which pockets of shock-produced glass were found to contain trapped Martian atmosphere (Photograph courtesy of NASA-JSC, Houston.)

EET A79001 contains numerous dark patches of glass, distributed throughout the meteorite in pockets. This glass was made by localized melting during the event when the rock was thrown off its parent's surface by an impact. When the glass is analysed in the laboratory, trapped gases are released. These gases have the same composition as Mars' atmosphere (as measured by Viking), which demonstrate the meteorite's Martian origin

More recently, another Antarctic Martian meteorite (ALH 84001) has been studied. It contains patches of orange carbonates (Plate 2) throughout its entire 1.9 kg mass. The carbon and oxygen in the carbonates indicate that the grains were produced at low temperatures from Martian atmospheric carbon dioxide dissolved in water, probably just below the surface of Mars, when water circulated through it. Scientists from NASA think they might have found evidence for fossilized Martian bacteria inside these carbonate patches, showing that life might have existed on another planet. By studying meteorites from Mars, we can learn about events that have taken place in the past on our neighbouring planet, when it had a thicker atmosphere and could support running water, even though the surface of the planet now seems to be dry.

Conclusions

Meteorites are a diverse set of extraterrestrial materials, representing planetary and Solar System material in its many forms. By studying meteorites, we can study processes that have taken place as our Solar System evolved. We can also learn about the evolution of other stars that contributed to our solar neighbourhood. Without meteorites and comets, it is likely that life would not have evolved on Earth; studying Martian meteorites might possibly allow us to study the primitive beginnings of life on another planet. But meteorites are not only associated with the seeds of life: they are also instrumental in influencing evolutionary pathways, as a consequence of catastrophic impact and associated environmental changes.

The study of meteorites is a constant reminder of our mortality, and can best be summed up by the sober phrase that warns us: 'dust to dust....'

Acknowledgements

I would like to thank Matthew Genge, Robert Hutchison and Ian Wright for reading the manuscript in its several manifestations, and improving the text with their comments.

MONICA M. GRADY

Born 1958, received an honours degree in chemistry and geology from the University of Durham in 1979, then went on to complete a PhD on carbon in stony meteorites at the University of Cambridge in 1982. She has continued to specialise in the study of meteorites, carrying out research at Cambridge, then the Open University, prior to joining the Natural History Museum in 1991, where as Senior Scientist in the Department of Mineralogy, she carries out research on and curates the national collection of meteorites. She was fortunate to be part of the 1988/89 US expedition to Antarctica to collect meteorites. Her particular research interests are in the fields of carbon and nitrogen stable isotope geochemistry of Martian meteorites, interstellar components in meteorites and micrometeorites.

Television beyond
the millenium

WILL WYATT

Introduction

In dutifully observing the rules relayed to me for my appearance before
you tonight, I don't know whether to be encouraged or intimidated by
the knowledge that they—the discourses and indeed the rules—have an
unbroken history going back to 1826. I may perhaps, as a non-scientist,
take comfort in the knowledge that previous audiences have sat through
presentations on Dr Marshall Hall's reflex function of the spinal marrow,
the manufacture of pens from quills and steel illustrated by modern
machinery, and the condition and ventilation of the coal mine goat.

But then, they didn't have television in those days ...

Television is what I propose to discuss, and it's possible, without
exaggeration, to say that many of the essential elements of the technology of TV were first demonstrated at these discourses. The first electrical flash photography in 1851, the announcement of the discovery of the
electron in 1897, and on 13 March 1882, Eadweard Muybridge demon-
strating 'the employment of automatic apparatus for the purpose of
obtaining a regulated succession of photographic exposures.'

These were the first moving pictures that most there present had ever
seen, and in the dignified interest with which they were received there
is little to suggest a future for the medium beyond scientific research
and exploration. A hint, however, of something more is contained in the
guest list for that night; the Prince and Princess of Wales, Gladstone,
Huxley, and Tennyson. If Muybridge is the stepfather of cinemato-
graphy, does that make them the forefathers of the couch potato?

Television craft and its technology are inextricably linked and I hope
to show where they have brought each other and to speculate a little on
where they may be heading—from an uncompromising single channel

presided over by the autocratic paternalism of one man, to a future where control may be not only impossible and perhaps unwelcome—but where responsibility, excellence and creative vision will always be required.

There are some ideas that are so great that, once somebody has thought of them, an unstoppable snowball of research and experiment begins to roll to will them into reality. So it was with TV. The word and idea of TV entered the world long before the technology brought it into being.

The word 'television' first appeared in 1900 but the theory was around from 1875. Scanning imagery—still the basis of TV was invented before World War I on the theory that light can generate electricity, so a lit image can be converted into electrical signals and sent via radio to be reconstructed elsewhere. The early experiments suffered because they couldn't generate signals strong enough, but then valves came along, and by the early 1920s there were several competing potential systems.

John Logie Baird, the name always associated with TV, started his experiments with TV in 1923. In 1925 he gave his first demonstration of 'the transmission and reception of moving images' using his spinning disc system. This was the original Baird televisor: essentially, the system shined a light through a disc perforated with a spiral of holes which stored the lit fragments of the image in front of it and reconstructed them as a series of vertical strips.

At the time the BBC was a private company owned by a consortium of radio manufacturers. Baird used their facilities informally to make his test but couldn't make any public announcement of it.

Things hardly improved for Baird in 1927 when the BBC was turned into a corporation. Lord Reith, its first Director General and a fellow Scot, was more than sceptical of this new fangled technology. In October 1928 Reith's assistant, Gladstone Murray, attended a demonstration and made the following report:

> The Baird system ... provides an interesting laboratory experiment ... The demonstration considered in terms of service might well be considered an insult to the intelligence of those invited to be present ... We have a primary duty to the listening public to do what we can to promulgate the truth and to prevent the excitement of false hopes.

Baird, however, was undaunted and was already advertising his receivers for sale, even though the only programming he offered were tests sent from his lab to his receiving station in North London. The cost of a set was £150—more than the price of a fancy car—but nevertheless some 'televisors' were sold. In September 1929, the BBC was reluctantly

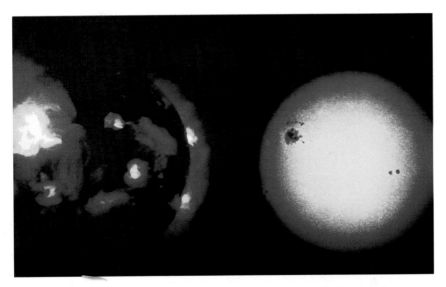

1. The sun in both visible light (*right*) and in X-rays (*left*). The dark areas visible (sunspots) are seen to be the seat of energy which manifests itself as X-rays in the solar corona. (Yokoh satellite.) (See p. 11.)

2. Patches of orange carbonates on the surface of martian meteorite ALH 84001. The field of view is about 0.5 mm across. The grains were produced below the surface of Mars when water circulated through it. (See p. 65.)

3. The historiated letter 'I' from *Genesis* in the Lucka bible. (See p. 159.)

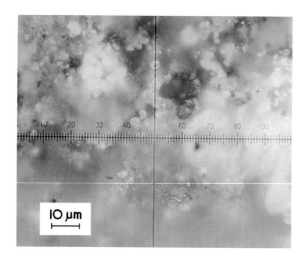

4. Magnified (× 1000) portion of the dark grey column depicted in a sixteenth century German manuscript showing the presence of at least seven pigments. (See p. 160.)

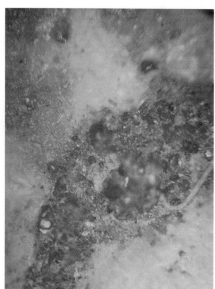

5. Photographs and microphotographs of elaborately decorated initials on a fifteenth century German manuscript MS Ger 4 (f 2r and f 28v) from the DMS Watson library. (See p. 161.)

6. Faces on the Byzantine/Syriac lectionary (upper f 188v, lower f 67v) blackened by degradation of white lead to black lead sulphide. (See p. 161.)

7. Partially reconstructed glazed bowl painted in black on a blue background (CF 1412). The scale section on the left is 2.5 cm long. (See p. 164.)

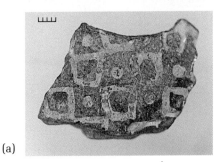

(a)

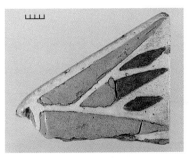

(b)

8. Polychromatic Egyptian faience samples from the Petrie Museum, (a) UC. 686 red pigment identified to be red ochre and (b) UC. 888 (lotus) yellow pigment identified to be lead(II) antimonate, $Pb_2Sb_2O_7$ (scale in mm). (See p. 164.)

persuaded to give him access to its one medium wave transmitter for experimental public broadcasts from his workshop, now in Long Acre. They were still keen to distance themselves:

> *In granting facilities for the experiment demonstration in which the public can, if they so desire take part, neither the Postmaster General nor the BBC accepts any responsibility for the quality of the transmission or for the results obtained.*

Broadcasts only happened on weekdays from 1100 to 1130 in the morning. Because of the limitations of the technology and his budget, most of them were simple solo performers of various sorts, but in 1930 Baird televised a play, *The man with a flower in his mouth*

1930 saw another first—the earliest recorded example of a published quote about TV's supposed pernicious effect on public values. And, what's more, it was in the BBC Yearbook:

> *Television might provide the most wonderful entertainment the world has ever seen, but might alternatively, if the control fell into the wrong hands, see all entertainment debased to the level of international millions or used for vilest propaganda.*

The reluctant entry of the BBC

In August 1932 the BBC took matters into its own hands by taking over production in a small studio in the basement of Broadcasting House.

This was still mechanical, or 'low definition', television using the spinning disc system, producing a flickering image with only 30 vertical lines, repeating at 12 and a half frames per second, tinted orange and, when viewed at the receiving end, measuring only 3 inches by one and a half. This experience may be familiar to those of you who have tried to download video clips from the Internet. The parallel is suggestive. Even though the pictures were extremely crude by modern standards, people at the time were surprised to find that the images were recognizable.

Once the superior editorial capacities of the BBC were brought to bear on it, content improved considerably. Though adaptations of great works of literature were still in the future, the high editorial purpose achieved a landmark when, late in 1932, they featured a seal which played the saxophone. (A triumph only excelled in 1979 by *Nationwide*'s skateboarding duck.) Naturally, the seal was considered something of a celebrity and arrived at the studio in style in the back seat of the engineers open tourer. The engineer's mistake was to escort this artiste in through the front door of the newly-built Broadcasting House. Memo writing was evidently already a high art in the Corporation, and on this occasion one was unleashed to enshrine a new internal regulation: 'Performing animals must in future use the goods entrance'.

Other problems were caused by the fact that the producers had been put in a studio also used by Henry Hall and his band for rehearsing. For a while they tried to fit round it, but eventually Hall was sent upstairs and the TV people were allowed to install a hard floor over the carpeting and a screen that hid the clock and the lavish, but inappropriate, decor of the room.

Things were not easy in front of the camera either. Let me introduce, Dee Barron. Dee is made up with the heavy blue-black, white, and yellow ochre make-up which emphasized the features of the early performers. The costumes and decor were also determined by technology; any large areas of single colour would weaken the signal and lead to fuzzy, smeary images. So the production staff had to design very graphical, decorated backdrops, costumes—and even masks. In case performers arrived inadequately patterned, a dressmaker was on call to sew on trim in emergencies.

The first night's broadcasts on 22 August 1932 were reviewed in *The Times* of the following day with decorous restraint: '*The broadcast was quite successful and one could easily recognise the artiste ... it should encourage a wider interest*'.

So the make up worked.

In November 1933 a 'caption machine' was devised to generate the first graphics: a mini-spot system for fixed titles and then a system which rotated a drum to position a selected card in front of the spot.

The audience at the time were called 'lookers in' or 'lookers' and a lot of them, presumably those who didn't have an extra £100 to spare, made their own sets from kits, including one sold by the *Daily Express*. Cross media ownership goes back a long way. And some of these do-it-yourself lookers in were very far afield: depending on the weather and time of year, people were getting pictures in Madeira and North Africa.

High definition and cathode ray technology

But all the enthusiasm could not overcome the primitive technology. Baird's scanning beam did not give off enough light, the amplifiers were not powerful enough to transmit a strong signal, and the device in the set for reproducing these images, was inadequate. The solution again derived from an idea dating back to 1908, which proposed the use of a cathode ray tube, invented 11 years earlier.

The first demonstration of cathode ray reception was given again by Baird in 1933. It was still small—4 by 3 inches—greenish tinted and only visible in a darkened room. A simple cathode ray tube was a vacuum tube with a phosphorescent screen at the front. A cathode at the back fired a stream of electrons across the tube to the phosphor coating which could be seen to glow green when the lights were turned down.

The early tubes didn't have 'charge storage' to keep the dots on the screen long enough to make the illusion of a whole continuous image, but the British government had begun to develop radar and they needed cathode ray tubes for that.

In 1931 money had been given to the EMI/Marconi Company to start a laboratory in order to develop better tubes, and in January 1935 the government recommended the initiation of a 'high definition' TV service. But Baird, not knowing about this deal, had continued to improve his system. By this time he had got his definition up from 30 to 240 lines and his image size up to $14\frac{1}{2} \times 10$ inches—but EMI's was 405 lines and scanned more rapidly, producing less flicker.

This EMI camera was also lighter and more mobile and flexible. Essentially, a mechanical system was always going to lose out to electronic technology, but Baird's persistence and his status as a pioneer, earned him the right to compete. So the service was launched with two competing technologies—used on alternate days.

The site chosen for these was the roof of Alexandra Palace, a nineteenth century exhibition centre some 350 feet above sea level in North London. The staff who moved in there in 1936 to start up the new system initially thought they had 4 months to familiarize themselves before they went on the air. However, the move coincided with the annual Radio Olympia show and a panic message arrived that nobody was buying TVs. This was not unreasonable because there was nothing to watch on them. So, the 4 months became 9 days, and they got on the air in time to broadcast to Radio Olympia twice a day for 10 days from 26 August.

Early days at Alexandra Palace

The service was officially inaugurated by the postmaster general at 3 p.m. on 2 November 1936. The short time span of TV's history was brought home to me when I sat next to Dallas Bower, the producer of the main programme of that first day, a half hour variety show, at a commemorative lunch a few weeks ago.

Reith, nothing if not consistent, was still grumbling and still contemptuous: his diary for 2 November reads: '*To Alexandra Palace for the television opening. I had declined to be televised or to take any part. It was a ridiculous affair and I was infuriated by the nigger stuff they put out. Left early ...'* But later he began watching despite himself—and his worst fears were realized. He wrote: '*Television is an awful snare...*'

The advantages of the Marconi system soon became very obvious and on 30 January 1937 the Baird transmissions ended. At that time broadcasting still only lasted 2 hours each day and not at all on Sundays. From 1937 it jumped to 16 hours a day. This sudden jump in hours led many people to wonder how all that air-time could possibly be filled. Among them was Jean Bartlett, an assistant producer from the early Baird days, who wrote: '*There still remain vast gulfs of time to be filled by programmes. Technical developments to the lay reader seem nothing short of miraculous. Can the imagination of the programme builders keep pace with them?*'

Technical constraints were still hefty enough to put quite a brake on that imagination. The cameras were still big, and without zoom lenses the only way to change the size of the shot was to change the lens on a rotating turret each time. So programmes still tended towards those where the action, if there was any, happened directly in front of the camera. So into the studio were brought variety acts straight from the London hotels and theatres where they were currently playing. Among them in the early days was a very young Margot Fonteyn, an unknown dancer at the VicWells company whose impresario Lillan Bailiss, had offered the services of as many of her singers and dancers as the BBC required.

One of the most popular shows was *Picture Page*, which in the way of new technologies modelled itself on a familiar one—the telephone exchange 'switching' viewers through to the next act.

Mickey Mouse Cartoons were also popular. The relationship between TV and the cinema was for long a delicate one, with the film business wary of the rival medium which threatened to cannibalize its audiences which of course it did, yet slowly realizing that this was a new and separate market for their productions. This suspicion started early and the only company that released to the BBC its films to show was Walt Disney, who let them have as many cartoons as they liked for nothing. The Disney Company today takes a rather different view.

Outside broadcasts and first sports coverage

From the beginning, teams of production and technical staff had been eager to get out of the confines of the studio, and had experimented with the idea of an 'outside broadcast' (OB). In the early days this simply consisted of ever longer lengths of cable, which could extend from the studio out of the building into Alexandra Park. Within a year, however,

TV engineers were building the first lorry equipped with mobile recording and transmitting equipment. It was called an OB unit and it was a major step in the development of modern TV. When it was only 4 days out of the factory, the unit was used for the first ever OB—the coronation in 1937 of George VI. There was no producer, the cameras and lorry load of equipment, was manned by the R&D team who had put the unit together. It was the chief engineer who stood looking down on the procession and suggested to the people in the lorry when they should cut to the pictures of a different camera

Many people bought their first set for the live coverage of the Coronation; viewing figures were estimated at 10 000—a huge audience at the time, but still far, far fewer than the people actually present on the ground.

It was the birth of the OB units that brought into being what is still one of the great benefits of TV—bringing an event of national importance to the whole nation, allowing viewers to share experiences and build a common stock of shared memories, as part of the national culture.

It wasn't long before the BBC OB lorries became a familiar sight at major sporting events. By 1937 the OB unit, with a radio link mounted on a lorry, was first at Wimbledon. In 1938, it was at Wembley for the Cup Final, by the Thames for the Boat Race and at Epsom for the Derby. Events which previously had only been recorded on film were now broadcast as they happened. When Chamberlain returned from Munich in 1938 the OB unit was on the tarmac at the airport to cover the event; the first example of electronic news gathering. War was looming and the BBC TV was about to be put on ice.

The war and TV shuts down

On 1 September 1939, broadcasts ceased with the immortal words 'I tink I go home' from one, Mickey Mouse. War was imminent and TV could not continue because the broadcast frequencies were needed for that other new development—radar.

It was nearly 7 years, on 8 June 1946, before TV came back with a bang for the OB of the Victory Parade—this time seen by 100 000 people.

After the war—birth of TV news

The 8th of June 1946 also saw the first broadcast of TV news. The BBC's radio news department was extremely suspicious of its frivolous sibling and thought TV lacked the gravitas for its weighty subject matters.

The radio people reluctantly agreed to a deal whereby they'd supply the content—including the news readers—and TV would supply the pictures. However, the cameras weren't allowed to show the news readers for fear their facial expressions might betray some bias, so the public was treated to an endless series of caption cards and maps.

Things improved, but not much, when the policy was changed. Now the people reading the news were visible, but they were deliberately chosen to lack personality and as they were also reading from scripts without Telepromters it remained a fairly arid viewing experience. Worse, it made the news readers look shifty rather than solemn, suspicious rather than serious, and the tabloids promptly dubbed them 'the Guilty Men'.

The 1950s and arrival of ITV

Technology continued to drive programming: 1950 saw the first OB from abroad—a link (by microwave) from Calais. I remember watching it. TV also continued to reach out to new audiences. The Coronation of our present Queen in 1953 was the first broadcast event which had a bigger TV audience than a radio one. It was a critical event in British TV history, because from this time, TV ceased to be a novelty and became a central part of Britain's creative and informational landscape.

It became literally part of the landscape, for the prices of sets were falling and by 1955 aerials were sprouting all over the rooftops of Britain. And there were now 13 transmitters with the potential to reach 92% of the population.

The BBC no longer had a monopoly; there was a new kid on the block—ITV, a rival commercially funded channel. So began the competition for audiences. ITV brought new quiz shows and filmed series, and its news arm, ITN, was a personality driven affair from the start with Chris Chataway and Robin Day, whose complete lack of deference on screen knocked even world leaders off their pedestals.

The competition led to a number of new kinds of popular BBC programmes: *Panorama, Tonight, Your life in their hands*, and John Freeman's *Face to face*. These programmes made use of studios and OBs as before, but in addition there was a new lighter 16 mm film camera easier to carry, easier to move altogether a more flexible tool than the bulkier 35 mm cameras.

Producers were soon making short documentary films for TV, and a new kind of TV reporter was born. People like Fyfe Robertson and Alan Whicker reported on the everyday events from up and down the country, bringing the experiences of British people to a TV audience. They presented stories which were sometimes serious, sometimes funny, but almost always reflecting a unique angle on some part of British life.

Later, as film makers wanted to capture on film yet more intimate portraits of life in Britain, reporter-less, cinema verite films and the concept of the fly on the wall documentary film was realized. I say films, because until 1958 there had been no videotape. TV was either live broadcast or film which had to be shot, processed, physically cut together with razor blades and sellotape, and finally taken to the transmitting station to be run from a telecine machine.

The reason we have a record of the early transmissions and some of the other historic events is that they were also covered on film, or film cameras were turned on to the TV screen to make a record of the transmission. When plays were repeated, the cast had to come back to the studio a few days later and do it all over again.

But all that changed in 1958 when the Ampex company introduced a system whereby an electronic signal from a camera could be stored on magnetic tape. The reels were heavy, the tape itself 2″ wide, so the recording equipment was also big and cumbersome but at least it meant that TV programmes could be stored for later transmission and/or for repeats—and for sales abroad. From then on the BBC vaults became a prized resource.

Miniaturization and simplification of technology

It has become a familiar constant with electronic technology that things always get smaller and usually cheaper. It's true of everything from calculators to mobile phones, and it's certainly true of TV cameras. Smaller and especially lighter cameras are obviously conducive to mobility and flexibility, both of which are good for most purposes. But the need for a cathode ray tube put a physical limit on how small they could be.

The breakthrough arrived with the charge coupled devices (CCD). Instead of a tube, the CCD cameras encode the information electronically, on three silicon chips. So now, the size of the camera is limited only by the size of the lens and the size of the tape it's recorded on to.

Analogue video tape degrades fairly sharply as it gets smaller and the signal has to be squeezed on to a smaller area. Even the best Hi8 camera with the best lenses produced a pretty rough image—certainly not the equivalent of professional modern Beta cameras, still less film. But these small cameras could be mounted in unlikely places and capture events which would otherwise be impossible.

The medical profession took advantage of miniature cameras, and this enabled them, their patients, and indeed TV viewers to explore inner space.

Later, when more traditional TV cameras became smaller, cheaper and more user-friendly new programme ideas emerged, notably *Video diaries* which became a regular series where people captured their own experiences, without any need for a technical crew.

But recently videotape has gone digital and the results of that are pretty spectacular. This small digital video camera produces an image about the same quality as a Beta camera which is much bigger and very much heavier.

What makes this revolutionary in the history of TV is that for the first time, we have a high-quality camera that can be used not only by specialist camera teams but by anyone who can afford just £2500. This camera was made for the domestic market. But BBC producers are already embracing this new technology enabling them to make some programmes at a fraction of the cost of traditional methods.

These small cameras also provide programme makers with all kinds of new possibilities. On the one hand we have the capability of getting quite extraordinary views of events but on the other hand the possibility of filming secretly becomes easy and more tempting. As with genetic engineering and other new technologies we have to be wise to the implications and aware of the uses and potential misuses of it.

A recent unlikely location is a cricket stump with a camera, which consists only of a lens and a circuit board that encodes the information. It is linked by microwave to a recorder, and the camera itself can go almost anywhere. The Australians came up with the idea of putting a camera in the stump for cricket coverage, but the BBC improved on the technique by replacing hollow wooden stumps (liable to splinter disastrously) with more or less shatterproof fibreglass. Naturally, this iconoclastic step was taken in consultation with the cricket authorities and the skills of our highly trained technicians were able to produce a stump indistinguishable from its fellows—even to the England wicket keeper. The one outstanding problem remains the tendency of umpires to replace the stump facing the wrong way. Work, however, is underway on a remote steering device that should solve that too.

This year a BBC TV camera travelled to an even more unlikely location—inside the pith helmet of a Marine during the Beating Retreat in June. The helmet happened to come with a convenient badge hole and a judicious dab of gold felt tip completed the disguise. The BBC is again covering the bob sleigh event for the Winter Olympics in early 1998, so you can expect some vertiginous shots by then.

Editing and the action replay

But to go back to the first video recording on tape: once it had become possible, all kinds of production opportunities presented themselves. Richard Dimbleby had demonstrated the notion of immediate playback for the first time on air in *Panorama*.

But of equal significance, the editing of studio and OB programmes was now possible. This in turn led to new production techniques: we saw the introduction of pre-recorded material, added as inserts to otherwise live programmes. Plays could not just be recorded but performed and recorded out of sequence, and the birth on videotape of the 'retake' or 'shall we do that just one more time?'

At first editing was a clumsy technique. The 2″ tapes had to be physically cut with a razor blade, and stuck back together at the appropriate junction often with a visible bump on screen. But the invention of electronic editing transformed the process from being a primitive safety device—a way of correcting mistakes—to a tool of production, which changed the way that programmes could be made.

If you look at news and current affairs programmes today, they are built entirely using this technology: some reports are recorded hours before, some minutes before, others are edited as the programme is being broadcast, to be slipped in later in the bulletin. And the running order, the sequence in which the items are shown, can, and indeed frequently does, change as the programme progresses. Modern digital editing allows us this flexibility.

In sports production, video recording provided the means to offer edited highlights as soon as an event was over. It provided the replay. They have developed to become part of the narrative of televized sports and in so doing, challenged the authority of the referees or, as in cricket now, provided them with a new aid. The action replay is, of course, under the control of a computer. And it was computer technology which was used to create dramatic advances in graphic design. Today, graphic designers have all but abandoned their pens, pencils, paper, and ink. Long gone are the days when someone was employed to change stencilled caption cards by hand. The computer can capture still frames, combine pictures, alter existing images, add text—with a software programme ironically named 'Paint Box'. The designers used this new found chip power to invent a new art form—the opening title sequence.

The front titles of the Royal Institution Christmas Lectures, use the latest computer technology, the Hal, a successor of Harry. The explosion of the cracker was created by photographing 30 still frames of cotton wool shaped progressively by the designer to create the effect of an explosion, when played back in real time.

There is no doubt that computer technology has given rise to a new industry in television: post-production. This is the packaging, adding new dimensions and a new sophistication, to recorded programmes. It's a sophistication which is still expanding as we move from complex animation to the realm of virtual studios.

But now, let me now move on to look at the changes outside of the production experience, which have also dramatically altered TV viewing for the audience.

Standards

When TV was half as old as it is today—On 21 April 1964 to be precise, the technical quality of broadcast TV went up a notch with the launch of BBC2, which broadcast on UHF frequency with 625 horizontal lines on the screen instead of 405. To receive it, the public had to be persuaded to buy a new dual standard set. The software—a greater choice of programmes—was the bait to get the public to buy the hardware—the receiver and a new aerial which became another accessory adorning Britain's skyline.

After buying and fitting this new apparatus the public waited expectantly in front of their new sets on the appointed evening of the launch. They were disappointed. A failure of the existing technology—a huge power cut in London obliterated the evening's viewing, so BBC 2 began with the first ever *Play School* the following morning. However, after the initial shock, the robust opinion of the television manufacturers was that the spectacular calamity of the opening night was the best publicity they could possibly have had.

In 1967 the lure of TV was strengthened by the introduction of colour, initially only on BBC2, but then in 1969 also on BBC1 and ITV. Colour had its first great showcase with the 1969 investiture of the Prince of Wales, which used just about every colour camera in the country. It was watched by $16\frac{1}{2}$ million people.

TV has an unparalleled role in uniting the population through mass experience and the internationalization of that experience was furthered by satellite transmission. It meant that events around the world could be broadcast live from even the most remote locations.

Sputnik 1 launched by the USSR in 1957 heralded a new era in communications. And in April 1965 Intelsat 1 or Early Bird, as it was known, became the first satellite to transmit television traffic on a commercial basis. Before that experimental systems had been launched and the most successful of these was Syncom 3 over the Pacific which was used for the Tokyo Olympics in 1964.

These could be seen live by satellite as far as California but had to be relayed by line and finally by aeroplane to people watching in Britain. By 1966 satellites were in place to link all five continents. The first big event to celebrate this, a monster programme, was *Our world* in 1967. 24 countries saw it live, hosted in Britain by Cliff Michelmore.

By 1968, the Mexico City Olympics were live, in colour, around the globe and in 1969 the world witnessed the furthest OB yet—from the moon.

Satellite technology has of course been the enabler, in this century, of a burgeoning of new TV channels. The existing distribution from terrestrially based transmitters was using most of the available frequencies. Distributing by satellite direct to peoples' homes via a special dish bypassed this frequency traffic jam. Distributing via cable does the same but the capital investment required to lay the cable is huge and here it has spread itself more slowly than in mainland Europe.

Both distribution routes have, by providing more choices, helped move the balance of power from broadcasters and producers to viewers. Easy access to these choices would be impossible without another piece of hardware—one of the most significant to date—hardware which has been helping to shift that balance of power.

This suitcase is the latest in satellite technology as used by BBC News. Within the past 10 years, news crews have been able to travel with a suitcase carrying the latest in satellite technology. The pictures shot and edited on location can be transmitted back to base, from wherever in the world they happen to be. And, of course, the physical material has no need to pass through national borders, so the control of the reports lies with the broadcaster. The Gulf War correspondents had two kits similar to this one, and for the first time some of the coverage of that war was instantaneous and no longer censored by the military.

The remote control

Transforming the way programmes are watched is this—the remote control.

No longer do you leave your seat, move to the set, select a channel, return to your seat, settle down to watch. And if you wish to change channel you have to do it all over again. Now, you can sample at will. Switch the moment you lose interest if you so choose. Watch two—children manage three—programmes at once. Graze the services available; nibble at the passing images. Within each household there are tales of 'the battle for the remote control.' Many of the losers buy new sets which they take to bedrooms, or kitchens, for viewing alone, a phenomenon hitherto common with radio listeners.

Viewing habits changed with the advent of the remote, but I can assure you that with the explosive changes in the pipeline, we ain't seen nothing yet.

So what are these changes and how will they affect television in the future? In this last section I want to outline very briefly some of the new technology which is now under development, and speculate about the future.

Digital—what is possible

Currently, whether TV is recorded on film, on analogue or digital tape, it is transmitted to your home as an analogue signal. Channel 5, is the last analogue terrestrial channel. In future, new channels will be broadcast digitally, whether by satellite, by transmitters, or by cable, SKY will launch digital channels probably by the end of the next year. At the same time the BBC plans to start broadcasting digital services—BBC1, BBC2 in Widescreen, and a 24-hour news channel. This is not just one huge change. It is the beginning of a series of huge changes.

We have been experimenting for some time. Widescreen formats will enable viewers to see movies in the ratio they were filmed and will bring a new dimension to sports and major events. I was there, when for the very first time, we transmitted analogue and digital widescreen at the same time: The first digital widescreen simulcast—it was of the Trooping of the Colour—in 1997.

Digital allows so much more: not only is there a dramatic improvement in sound and picture quality, but many, many more digital channels can occupy the space of one analogue channel. The zoochannel world will soon be with us. For the BBC this means that extra programming can be offered alongside the digital BBC1 and BBC2, but also extra information. This will allow viewers to engage with the programmes in new and satisfying ways, following the paths of their own curiosity and interest in a way that has never previously been possible.

Just as the programme at the theatre or opera brings to life a wide range of information about the performance, so we can use the digital capacity to bring to life new levels of information and interaction. Participation is up to the individual viewer. People who just want to watch the show can still do so but people who want more can guess the value of that piece on the *Antiques roadshow* or read the score of *Young musician* in parallel with the performances.

Parallel soundtracks can allow you to choose whether to receive the original version of a foreign film or a dubbed one. You've missed the beginning of a drama or a film. At the press of a button you can have the story so far. We can offer more chances to catch up with the most talked about programmes or your favourite dramas.

And with partners it is hoped to launch some pay channels using the BBC's ever growing archive, theming and presenting the material to appeal to particular groups of viewers. Other broadcasters will be planning their offerings and we shall certainly see 'near video on demand'— the same film starting at 10-minute intervals on a dozen or more parallel channels.

Before long we will have moved from a world in which all viewers receive their TV by analogue signal over the air from terrestrial masts to a world in which TV will be received from many different exclusive routes—digital satellite, digital terrestrial, digital cable, and later down the telephone wires.

It is not just the distinction between broadcaster and audiences that is blurring. Digital technology is also blurring the distinctions between different broadcast and communications media themselves. Digital transmission technology makes no distinction between video, audio text, graphics, and pictures. The nature of any digital recording is that every piece of information is simply converted into a string of ones and zeros for transmission to a receiver which reassembles only those parts of it that its user wants to see and hear.

These receivers of digital information will continue to be known as telephone, radios, TVs, and personal computers (PCs). But at their heart all will be computers. They will differ in shape, size, and processing power, and they might offer a variety of screens, speakers, and printers, but ultimately at the heart of every receiving device in the digital world will be computing power.

What the digital world means is that in time the lakes and rivers of analogue will become one vast digital ocean. TV programmes, once they become part of the on-line digital offer, will no longer be part of the TV offer but part of the universe of all digital shopping, banking, interactive entertainment, and information services.

This means that digital media will tend to shift the balance of power from broadcasters to the owners of rights—sports, films, entertainment, and drama. In the old environment, TV was the richest analogue medium and viewers were restricted to a few terrestrial channels.

Because the supply of programmes far exceeded the channels' limited capacity to show them, channel owners could pick and choose content that they broadcast and reward it on a cost-plus basis.

Digital technology is removing those channel capacity limits and soon high-quality content will be much rarer than the capacity to deliver it. Therefore, the rights' owners will be able to pick and choose the transmission media rather than vice versa.

TV over the last 50 years has shaped the way that we think. We have come to expect and accept neatly packaged programmes in linear form. This has given enormous influence to media owners. Digital technology will partly diminish such influence. Linear TV will remain the most powerful form of communication but in the next 3–5 years the digital age will begin to have more meaning in our lives.

The TV set itself is being given a run for its money. Although worldwide there are about three times as many TV sets in use as PCs, with annual sales approaching 70 million, sales of PCs will overtake those of TV sets in the next year, and the PC is fast reaching the point where it begins to rival the quality of a TV set.

This sounds like a paradise of choice for the viewer. But, in order to get all these services, viewers will have to invest in a set-top box; both for them to receive the digital signals and for suppliers to charge for their programmes and services. That set-top box is the gateway to the digital future. The standards in that box are proprietary. So the keeper of the gateway is in a powerful position.

In Britain it is a uniquely powerful position. That is because the owner of those standards—Rupert Murdoch—also owns the rights to the most attractive programming. Sport and movies are in his own words the 'battering ram' to drive take up of pay television.

All credit to him for having had the vision and the guts to get BSkyB where it is. That has begun to change the market for good. But what happens next—and I mean over the next few weeks—will shape the future of digital broadcasting in Britain. If the government gets it wrong one player may end up dominating the market through digital satellite.

Let me give you two examples of what this might mean. First, it will determine how far interactive services will develop. The power of the PC through the TV. Potentially a tremendous force to change our lives—whether in education, information, home shopping, or banking. But without access to the proprietary standards in the gateway, broadcasters and others cannot plan and develop these services. They will happen at the pace and in the manner determined by the dominant player.

Second, with an enormously greater range of channels and ancillary services, context becomes as important as content. Thus the role of the electronic programme guide will be crucial—the browsing mechanism that helps you find what you are looking for. Currently, we have the *Radio Times*, the *TV Times*, *TV Quick*, and others. The electronic programme guide could be a Sky programme guide or it could be a *Radio Times*/BBC programme guide, just as it could be a *Hello!* magazine programme guide. It will be a powerful tool to direct viewers towards some services and away from others. For the BBC and ITV to be simply a few channels 'brought to you by BSkyB' on their guide is a grim but real possibility. That is what might happen in a dominated digital market. What is more, it will be a smaller market. Several million people who want it and can afford it will take digital satellite. But that is only a fraction of the 22 million homes in Britain who could enter the digital age.

If the government gets it right—as we are arguing for—then there will be an open market. One in which there is genuine interoperability between delivery platforms and interconnection with different service providers. The consumer will get a wider choice of new services and programmes and a choice of delivery platforms—satellite, cable, and terrestrial television. We intend to be on all these platforms. We hope that regulation helps ensure those platforms are there for us to be on.

What we don't know about the future is the impact that regulation will have on the way that these markets develop. But we are arguing for it to help the market develop not hinder it. We also don't know how the public may react. They are increasingly resentful at being forced to make technology choices to consume entertainment, and they may wait longer than anyone imagines in upgrading to the digital age. Different countries will upgrade to digital at different times, and will leap-frog each other in terms of their systems and platforms.

Broadcasters will only be meaningful if, like newspapers, they manage to bring trust, attitude, quality, and other appealing characteristics to help the viewer and listener find their way through a far more confusing—but exciting—media world.

Although there will be generational changes eventually, the one certainty is that people's tastes and sensibilities will change far less than technology or regulation. As we have seen in every other industry from food retailing to home insurance, people put a real premium on brands that they can trust because of the promise of the quality, reliability, and decency that comes with them.

In an increasingly crowded, noisy room, a friendly trusted face will become more and more valuable. The BBC will be enabled by technology not only to be a producer of much of the best material in the world, but to be a link, a broker if you like, between creators and audiences from all over the world who share our sensibilities and our values.

Finale

In this august environment, constrained by the rigours of scientific method in the search for truth, I dare not stray further into the murky waters of crystal gazing and prognostication. I've talked for long enough and as you'll have gathered my colleagues and I have quite a lot of work to do.

WILL WYATT

Born 1942, won a History Scholarship to Emmanuel College, Cambridge. He trained as a reporter on the Sheffield Morning Telegraph before joining BBC Radio News as a sub-editor in 1965. He moved to television production in 1968, working on programmes such as *Late Night Line-Up*, *The Book Programme*, and *B. Traven: A Mystery Solved*. He became Head of Presentation Programmes in 1977 and later Head of Documentary Features, responsible for such programmes as *40 Minutes*, *Around the World in 80 Days*, *Children in Need*, and *Rough Justice*. Appointed Managing Director, Network Television, in 1991, he has this year been appointed to the new post of Chief Executive, BBC Broadcast, part of a major BBC reorganization designed to meet the challenges of the digital age. He is a Governor of the London Institute and ex-Govener of the National Film and Television School. His book *The Man Who Was B. Traven* was published in 1980. He is a Fellow of the Royal Television Society.

Molecular information processing: Will it happen?

PETER DAY

Preamble: brains and foresight

Molecular information processing is going on all around us. Molecular-based information processing has in fact been going on for millions of years in its natural form, through the brains, not just of human beings and their predecessors, but in all other higher (and perhaps even lower) living organisms. For all brains, however rudimentary in evolutionary terms, operate at the molecular (or more precisely supramolecular) level. Thus nature has given us an existence theorem and the question becomes a different one: not 'Can information be stored and manipulated at the molecular level' but rather 'Can we ourselves design and manufacture artificial structures using molecules that will carry out these functions?' Sadly, at the present time the short answer to the second question is 'we don't know', but if that was all there was to be said about the matter there would be no point in continuing this exposition.

So I find myself in the rather unusual position, in tackling a subject that does not yet exist. Indeed, in some ways it is even a bit embarrassing because I am not known as a great fan of crystal ball gazing. Over the last two years the Department of Trade and Industry (now the civil service resting place for the Office of Science and Technology) has conducted a Technology Foresight exercise, and it is significant that no mention of molecular-based information processing appears in it. Perhaps this is one result of the very narrow subject base for the advice it sought: for example no chemist was invited to sit on the panel dealing with information technology (IT) and electronics, despite the fact that even the present generation of integrated circuit hardware is constructed by chemical means, using photolithography. Conversely, neither did any representative of the IT industry contribute to the deliberations of the Chemicals Panel. In fact, when the House of Commons Select Committee on Science and Technology took evidence about the procedure and effectiveness of the Foresight endeavour I have to confess to submitting (and defending under interrogation) a highly sceptical view on the subject. It is therefore starting from this sceptical and cautious base that I want to consider where the subject that has been called 'molecular electronics' has come from, what its present status is, and what a future information processing regime based on molecules might just possibly look like.

Let me begin by nailing my own colours firmly to the mast: I am a chemist, and a distinguishing feature of chemical science is to build new structures, that is, new arrangements of atoms into molecules and molecules into aggregates (which might be, but do not necessarily have to be, crystalline) in such a way as to create new properties and functions. As noted already, molecular chemistry is already contributing very significantly to the electronics industry as it exists today. Examples are the fabrication of silicon wafers, etching of fine structures on their surfaces using photoresists, deposition of thin films by chemical vapour deposition and, above all, displays made from liquid crystals, which we find in digital watches and laptop computers. However, that is not the focus of my present concern. I want to consider whether molecular assemblies themselves might be used as a means to store and process information. Put baldly like that it sounds like science fiction, but let us remind ourselves what has been happening in electronic data processing over the last 50 years, or indeed in other forms of processing for longer than that.

Decreasing size: from the abacus to (very large-scale integrated) VLSI circuits

The earliest mechanical form of data processing (if you leave aside counting on our fingers, which gives rise to the word 'digital') was the abacus. The simple form of this device, still widely used in bazaars throughout the Far East, still serves to illustrate a number of important points about computational functions that are shared by advanced electronic systems. First of all the information is stored and processed in 'bits', units symbolized by the balls that move on the wires of the frame. Second, the bits are manipulated serially, i.e. one after the other, rather than all at once. Third (a point that will come up again later), to move the bits takes a certain amount of energy, in this case muscle power, thus reinforcing the fact that each computational step consumes power.

The same considerations apply to more advance mechanical computers, which reached their apotheosis in Babbage's arithmetical engine, a device of gear wheels so elaborate that the handle could scarcely be cranked by a strong man. In more recent times mechanical calculators remained in use till some 30 years ago, overlapping chronologically with the first ones to use electrons.

The earliest devices to process information using electrons, such as the Colossus at Bletchley Park during the Second World War, were no smaller than Babbage's engine: indeed, the Colossus (which was well named) filled a whole room. Similarly, the individual components (valves) from which they were constructed were of similar dimensions to Babbage's gear wheels, i.e. a few centimetres. Furthermore, the programming and recording mechanisms remained mechanical, in the form of holes punched in cards or paper tape. Over the succeeding 50 years, however, the components of information processors have grown inexorably smaller, at the rate of one order of magnitude every 10 years. Valves were replaced by transistors, so that the electrons no longer circulated in a vacuum but through the conduction band formed from the overlapping atomic orbitals in a solid. Single transistors, linked by wires, gave way to integrated and then to VLSI circuits etched by elaborate sequences of chemical operations on to the surface of a silicon single crystal. Figure 1 gives a pictorial indication of the orders of magnitude that have been traversed.

Nowadays, the individual features found in commercially available personal computers are less than 1 μm (one-millionth of a metre) across and compare with the size of a virus. Enormously complicated patterns are achieved routinely by photolithography (Fig. 2) or etching with finely focused electron beams. Astonishingly, this evolution in the size of electronic computing elements, or features, follows an exact linear relationship with time, over quite a long period (Fig. 3). It has been called Moore's law, after the founder of the integrated circuit manufacturing company, Intel. In fact the relation shown in Fig. 3 turns out not to be the result of some hidden law of nature, but actually results from economics: the cost of building a new plant to produce microprocessors scales with their feature size, and currently it stands in the region of $1 billion, which explains why chip manufacturing throughout the world is concentrated in so few hands.

Given the vast size of the investment in time, effort, ingenuity (and hence money) to make microprocessor features ever smaller and smaller, it is legitimate to ask why all this trouble is being taken. Clearly, small size is an advantage in itself: we can carry around computing power in a laptop equivalent to the mainframe machines of 20 years ago. Individual

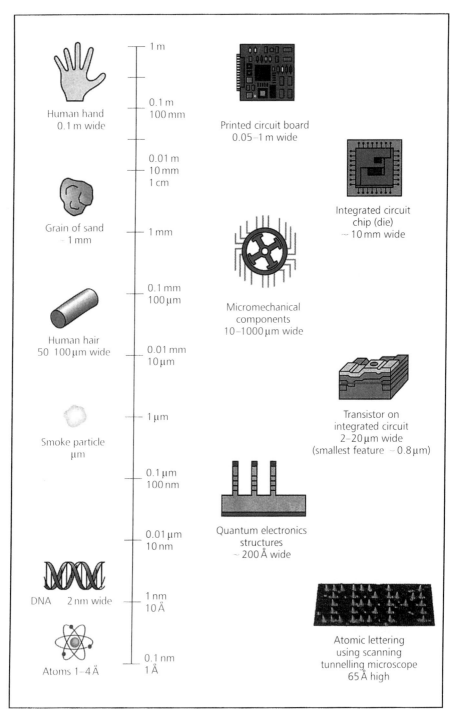

Fig. 1 A pictorial representation of the sizes of information storage media. Reproduced with permission from *Chem. Brit.*, **32**, 29–31 (1996).

Fig. 2 A modern microprocessor. Reproduced by courtesy of Sharp Electronics.

switching processes also take less power. Valve-based computers had huge power consumption but the laptop only needs a rechargeable battery. (Incidentally the latter is itself a marvel of solid state chemistry,

Peter Day

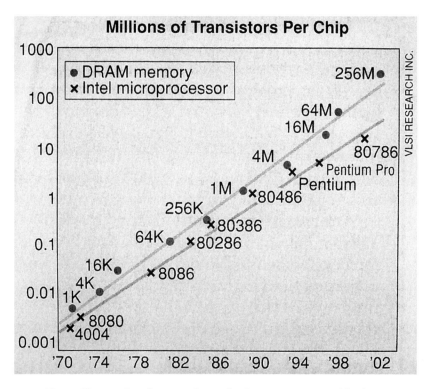

Fig. 3 Decreasing feature sizes of microprocessors with time.
Reproduced by permission from *Science*, **274**, 1834 (1996).

but that is a story for another time.) In view of these advantages to minia-turization, can we imagine any limit to the process, apart from human ingenuity and cost? Well, unfortunately, yes. There are three so-called 'fundamental' limits, set by the laws of nature. The first is set by the sta-tistics of the switching process in a binary system (i.e. on or off), which translates into thermodynamics via entropy. The second arises from the dictates of quantum mechanics, embodied in Heisenberg's uncertainty principle. The third, which is the most important one in practice is deter-mined by the heat dissipated by each switching event. As we saw with the abacus and Babbage's engine, it takes energy to throw a switch, and it is no accident that the largest volume inside the central processor unit of a modern supercomputer consists of pipes filled with refrigerating fluid! Altogether the three limits can be displayed on a single plot of power dissipation against switching speed (Fig. 4).

The final consideration brings me at last to the central point. Single chemical events, such as the reaction of a molecule catalysed by an enzyme, consume far less energy than the present generation of semi-conductor-based electronic switches (roughly 10^2 kT compared with

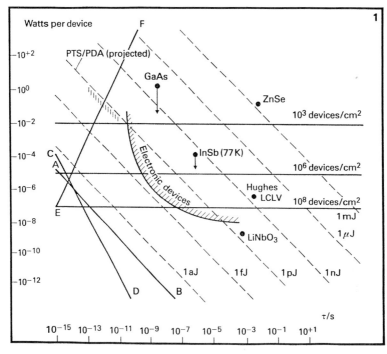

Fig. 4 The fundamental limits of computing speed and power dissipation. Reproduced by permission from *Chem. Brit.*, **26**, 52–4 (1990).

10^{10} kT). In principle, therefore, substantial advantages might follow from exploiting molecular chemical events as elementary information processing. If that is the case, several questions need to be addressed before we could contemplate any attempt to realize the potential of the new approach. First, is it going to be possible to make complex ordered structures from molecular components analogous to those of microelectronic circuits such as the ones in Fig. 3? Second, what are the kinds of physical or chemical processes that could be envisaged to store and process the information? And third (perhaps the hardest question) how could we get the signal in and out? Or, put another way, can we interrogate and detect the states of single molecules? The remainder of this account will be devoted to providing some answers to these questions.

Ordered molecular arrays: chemistry at work

Building structures out of atomic or molecular building blocks that are ordered over a long range in one, two, or three dimensions are at the very heart of the science of solid-state chemistry. For example, arrays of molecules forming domains with different orientations are shown in Fig. 5, where the components in question are long chains of carbon atoms (n-alkanes). Note the great, although superficial, similarity between this structure and that of Fig. 3. One of the most powerful methods of engineering ordered molecular arrays exploits the affinity of hydrophobic molecules, such as the alkanes above, for other hydrophobic ones, and of hydrophilic (literally 'water loving') for hydrophilic. It is termed the Langmuir–Blodgett method, and as it has been expounded in detail in Friday Evening Discourses I will not give more than the briefest of mentions here. This approach to making thin films one molecule thick on the surface of water is said to have had its origin in an experiment carried out by Benjamin Franklin, who poured a small quantity of oil on to the Serpentine pond in Hyde Park and, watching it spread out across the entire surface, was able to make a simple calculation of its thickness from the volume of the oil that he had poured out.

Apart from such rough and ready arithmetic, there are nowadays many ways of verifying that such films really are only one molecule thick. For example, one can build up multilayers by dipping a glass slide repeatedly in and out of a trough of water, which has such a

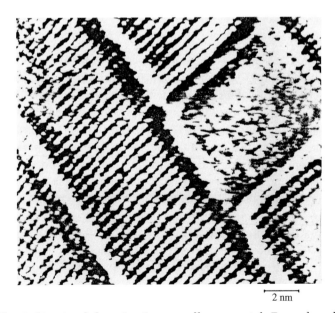

2 nm

Fig. 5 Structural domains in an n-alkane crystal. Reproduced by permission from *Atomic and nanoscale modification of materials*, P. Avonis (ed.), Kluwer Academic Publishers, Dordrecht, 1993, p. 263.

monomolecular film on its surface. If the molecular chains are terminated by hydrophilic groups, they stick in the water, and are transferred
to the similarly hydrophilic silicate surface of the glass. Thereafter, successive dippings produce layers on the glass with successive hydrophilic–
hydrophilic and hydrophobic–hydrophobic orientations (Fig. 6). How
do we know that single layers are being added at each stage? If the molecules each contain a light absorbing centre we can measure the absorption increase after each dipping: the increment turns out to be constant.
Even better, atomic scale resolution in electron microscopy permits a
direct view of the aligned molecules, as seen in Fig. 7.

Apart from these somewhat artificial methods of producing thin films of
molecules, three-dimensional architectural assemblies of molecules are
produced in wondrous variety by chemical synthesis. Figure 8 brings
together a few examples culled almost at random from the literature of the
last few years. Such elaborate constructions are not put together by what
has (somewhat disparagingly) been called 'engineering methods', i.e.
piece by piece assembly, but by the much more powerful method of
'self assembly'. The latter reaches its apotheosis in biology, where the
three-dimensional structures of enzymes are completely defined by the
sequence of amino acids forming the protein chain (the so-called primary
structure). Hydrogen bonding coils the primary structure into the α-helix
(secondary structure), which in turn coils into the final shape through the
medium of charge, and both hydrophobic and hydrophilic interactions.

An example of a low-dimensional magnetic lattice, which self-
assembles merely by mixing the ingredients in aqueous solution, is the
phosphonate salt synthesized by Simon Carling at the Royal Institution
(Fig. 8c). The organic and inorganic components in the structure segre-
gate spontaneously into alternating layers. Further examples from our
own work in the Davy Faraday Research Laboratory are molecular layer
compounds showing the highly unusual property of superconductivity.

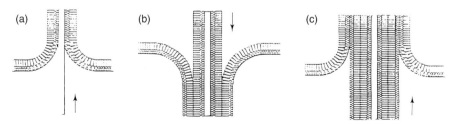

Fig. 6 Forming Langmuir-Blodgett films. Reproduced by permission from *Introduction of molecular electronics*, M.C. Petty,
M.R. Bryce, and D. Bloor (ed.), London, Edward Arnold, 1995,
p. 225.

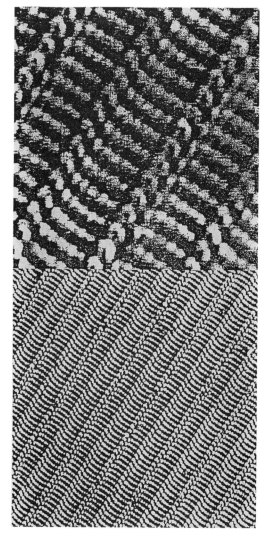

Fig. 7 Aligned molecules observed by electron microscopy. Reproduced by permission from *Adv. Mater.*, **8**, 903 (1996).

The compounds in question are so-called 'charge transfer salts', formed by oxidizing a heterocyclic aromatic molecule called bis(ethylenedithio)tetrathiafulvalene (or ET for short) and combining the resulting cations with inorganic anions. The oxidation is carried out most conveniently by electrochemical means and indeed, it has often occurred to me that the small glass cells used for growing crystals of these materials would be recognized quite easily by Sir Humphry Davy, should his ghost return to our laboratory!

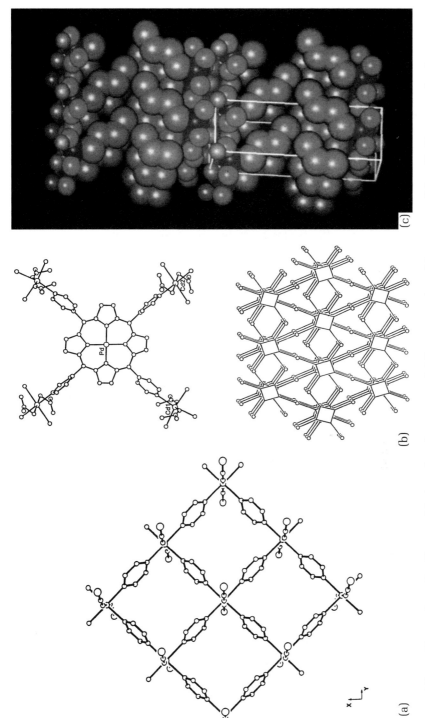

Fig. 8 Examples of self-assembled molecular arrays. (a) Iron pyrazine thiocyanate; (b) palladium porphyrin cadmium adduct; (c) manganese *n*-butylphosphonate.

The structures of one of the compounds prepared in this way is shown in Fig. 9. Again we see the organic part of the structure (ET) segregated from the inorganic part, $(H_2O)Fe(C_2O_4)_3.C_6H_5CN$. The compound is metallic, and becomes superconducting at low temperature. It is worth noticing in passing that we can now make superconductors containing water inside the crystal lattice, something that would have been thought quite extraordinary a few years ago. The compound in Fig. 9 also contains a high concentration of paramagnetic iron atoms, again something that would have been thought quite incompatible with superconductivity before, because magnetic moments are supposed to disrupt the Cooper pairs carrying the superconducting current.

Apart from self-assembly of molecular units into infinite arrays and networks, another development of recent years that has given a big impetus to the new field of nanotechnology and molecular electronics is the controlled preparation of clusters of atoms with defined sizes, and hence properties. So far as small metal particles are concerned, the origins of this topic lie far back in history, and actually have their origins in the Royal Institution. Michael Faraday made a colloidal solu-

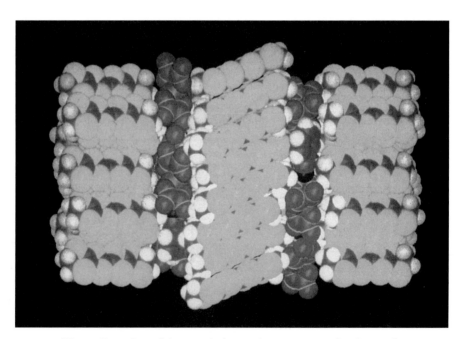

Fig. 9 Organic and inorganic layers in a superconducting molecular charge transfer salt synthesized in the Davy Faraday Research Laboratory of the Royal Institution.

tion of gold whose beautiful purple colour remains perfectly clear and uncoagulated to this day: it rests in a cupboard in the Director's study. Much more recently it has become possible to synthesize particles containing quite distinct numbers of atoms; for example, the one shown schematically in Fig. 10 contains 586 gold atoms. That seems a strange number, but you have to bear in mind that the packing of the atoms is the same as in bulk gold, that is, a hexagonal close-packed lattice of equal spheres. Starting from a single central atom, we can imagine it surrounded by 12 neighbours, then a further layer is added, and so on, generating a series of 'magic numbers' corresponding to the number of shells added. Finally, the particle is sheathed in an organic coating of alkanethiol molecules, whose sulphur atoms attach themselves to the outermost gold atoms. Similar particles can also be made from semiconductors such as CdS. and in that case the luminescence that takes place when electrons are excited across the band gap varies very markedly in

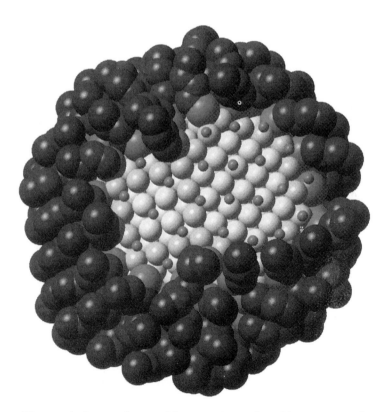

Fig. 10 A cluster of 586 gold atoms. Reproduced by courtesy of Professor D. Whetten, Georgia Institute of Technology.

frequency with the size of the particle. Particles about 20 Å in diameter emit blue light while those 40 Å emit red. Sizes in between emit every other colour in the visible spectrum. These are examples of what physicists now call 'quantum dots': electrons being confined in such small volumes that their energies are quantized.

Even more remarkable than the ability to manipulate the sizes of such tiny objects is the way in which it is now possible to move them around on a surface. Even atomic scale patterns have been made using the atomic force microscope: a famous early instance was writing the name of the company IBM in xenon atoms on a silicon surface. A circle of iron atoms on such a surface also acts as a 'corral' for electrons trapped inside it, whose wave-like energy states are reminiscent of the waves seen on the surface of the tea in a cup when it is shaken. Unfortunately, in many cases the atoms do not form very strong bonds with the surface, so they have to be placed and maintained at low temperatures. A significant recent exception, however, is the icosohedral carbon cluster molecule buckminsterfullerene (C_{60}): it turns out that they stick hard enough that they can be manipulated at room temperature, as, for example, into the S-shaped array illustrated in Fig. 11. From here it is just a short step back to the abacus mentioned earlier, and C_{60} molecules have actually been moved back and forth in a groove on a silicon surface to make a nano-scale model of this ancient counting device.

Mechanisms

If the answer to the question 'Can we make ordered atomic and molecular arrays with interesting electronic behaviour by chemical methods?' is a resounding 'yes', then what about the second question about the mechanisms that could be envisaged to store and process information? It was in this area that much of the early hype about molecular electronics gave the subject a bad name and led to justifiable scepticism. About 15 years ago an American theoretical physicist. Forrest Carter, began to speculate about the construction of gates and switches (necessary building blocks for information processing) out of molecular components. Many possible switching mechanisms were taken into account such as jumping of protons from one side to the other of unsymmetrical hydrogen bonds, electron transfer between donor and acceptor groups, and so on. Carter's ideas created a big stir and several conferences were held to discuss them. However, being a physicist (and a theorist at that) Carter had not really given much close thought to how such elaborate molecular struc-

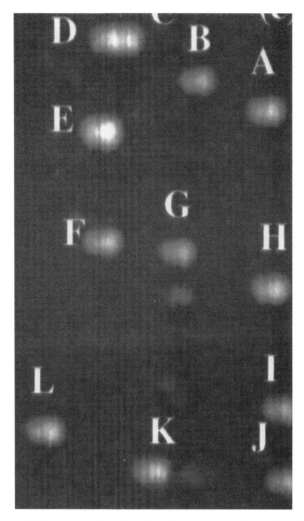

Fig. 11 Fullerene molecules placed in an array. Reproduced by permission from EPSRC Newsline, November 1996, p. 7.

tures as he conceived might be synthesized. His ideas therefore met with much scepticism, especially among chemists. An example of one of his proposals is shown in Fig. 12, which shows a hypothetical molecular wire with a gate attached to it.

Now in fact there is some practical basis behind the idea of a molecular wire, as such wires exist in nature in the form of conjugated polymers, of which the simplest prototype is polyacetylene, a chain of carbon atoms connected by alternating single and double bonds. If we

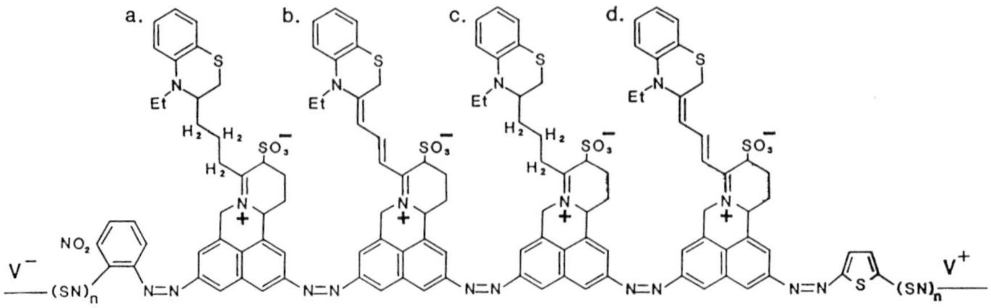

Fig. 12 A hypothetical molecular circuit gate (after Forrest Carter, 1984).

represent it in the *trans* configuration, it is clear that we can write two equivalent structures, in a fashion analogous to benzene:

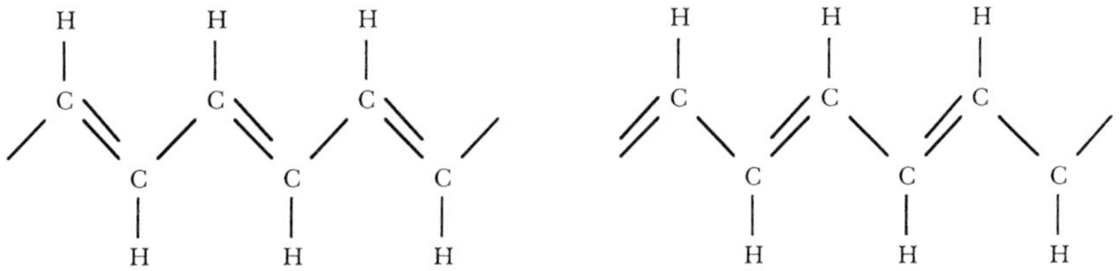

In an infinitely long chain the two would be indistinguishable but consider what happens if we reverse the conjugation in the middle of a chain:

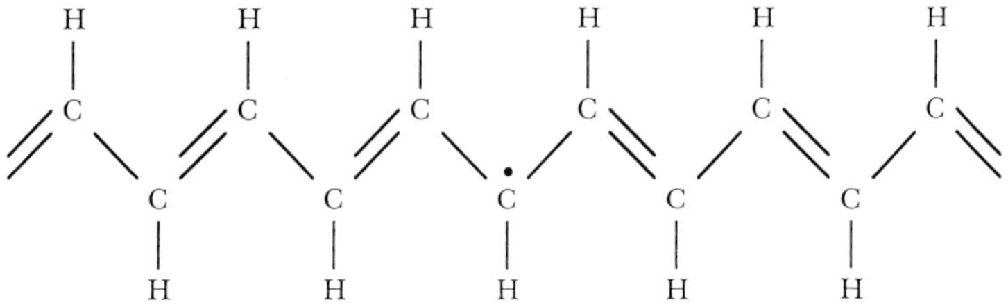

The dot signifies the presence of an unpaired electron separating the two parts of the chain where the alternation has different phases. By exchanging single and double bonds this singularity can move along the chain, taking the unpaired electron with it. It turns out the motion is analogous to that of a solitary wave, of which the best known example is the Severn Bore, the single tidal wave that sweeps up the River Severn every 24 hours. In the context of wave–particle duality, the solitary wave can be identified with a pseudo-particle called a soliton.

   Apart from solitons, several other developments in molecular science in the last few years lend hope of purely molecular data processing events. The heart of a transistor is a p–n junction, a region of semiconductor separating positively and negatively doped material. Such junctions act as rectifiers in the sense that on applying a potential in one direction current can be made to flow, but not in the reverse direction. A hypothetical molecular switch was the subject of a conceptual patent by Aviram and Ratner as long ago as 1974, but recently British scientists Ashwell and Sambles have realized the effect by building molecules that combine electron donating and accepting components into Langmuir–Blodgett films. A second, and quite different, approach has been taken by Stoddart, who has synthesized what he calls 'molecular shuttles'. A cyclic molecule forms a loop around a linear one or another much bigger cyclic one, and the loop moves from unit to unit up and down the chain. A method called dynamic proton nuclear magnetic resonance shows that in an example such as that of Fig. 13 the 'train' moves on the 'Circle Line' at about 300 turns per second. Two trains can be introduced using high pressure synthesis, when it turns out that they rotate at the same speed, never colliding! Other possibilities could involve the motion of ions, which is what happens when nerve impulses travel, or the transport of chemicals, which in biology is accomplished by neurotransmitters. All of these are chemical options for transmitting impulses on a molecular dimension, but for all of them there remains a fundamental question, the third on my list.

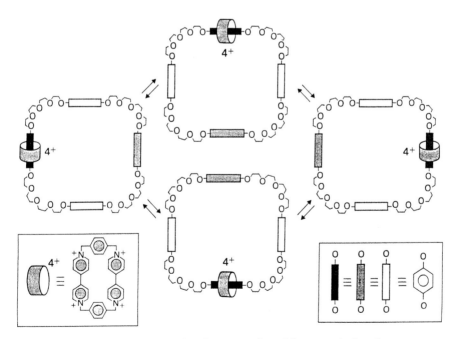

Fig. 13 A molecular shuttle. Reproduced by permission from *Chem. Brit.*

Addressing molecules

If the ultimate goal is to carry out data storage or processing at the truly molecular level, we have to face the question 'How do we get the signal in and out?' or, more starkly: 'Can we envisage ways of addressing the status of a single molecule?' At this juncture we have arrived at the point where analogies with conventional semiconductor electronics break down, despite the fact that (as we have already seen) it is already perfectly within the realm of possibility to manipulate molecules, or even single atoms by atomic force microscopy. It is also the point at which the conceptual models like Forrest Carter's fail, even if one could ever persuade a team of synthetic chemists to try and make one. The problem arises from a simple failure of imagination: Why should the architecture of a molecular information processor mimic that of one made from silicon?

Let us consider what tasks an elementary information processing unit has to perform, in an entirely abstract way. Such tasks are often set out in the form of a so-called 'truth table', summarizing the relation between the signal going in and the one coming out. Usually, the signal is assumed to be binary (i.e. either 0 or 1), but this is not essential. Two of the simplest of such gates, shown below, mimic the functions AND and OR:

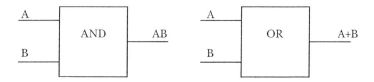

Now the rectangles are 'black boxes': what goes on inside them does not matter. From our point of view, are they single molecules (à la Forrest Carter) or something more complicated?

In conventional microelectronics the signal, consisting of a bunch of electrons, passes around the circuit from one element to another. Another way of expressing this is to say that the processing is serial. Now consider a set of individual elements that interact, not in a one-dimensional (serial) way but in two or three dimensions. In that case the state of any one element is determined by the number of its neighbours in the same state. Thus is born the notion of cooperativity, leading to bistable states when the whole set undergoes a transition from one state to another. Such phenomena have been used to store data on a much larger length scale for many years, for example in magnetic memory devices. The ultimate requirement is for the individual elements to be capable of existing in two states (charge, spin, vibrational: it doesn't matter), with a mechanism for each element to be sensitive to the state of its neighbours. One practical example of this kind, where the elements are transition metal ions bridged by organic molecules, already exists in the form of lattices in which metal ions change their spin states. Indeed, it is entering service with France Telecom as a means of reading phone cards.

Even in the realm of conventional microelectronics, structures are coming to resemble more closely the requirements of a molecular array. As the 'feature size' (jargon for the individual junctions and processing elements) becomes smaller and smaller, one arrives, willy nilly, at a situation where they all interact with one another, so that it is no longer possible to define a unique pathway carrying the electrons across the surface of the chip. Furthermore, to optimize the manufacturing yield and minimize the number of lithographic errors, designers have to make the units simpler and their arrangement more regular. Such a regular connected array, each node of which can exist in either of two states (on, off, or 0, 1) is an example of what has been called a 'two-state cellular automaton' (Fig. 14). The point here is that the entire lattice behaves as a single system. To store or process information you do not have to interrogate each element to find out its state: the information is stored in the form of a pattern. This idea forms the basis of

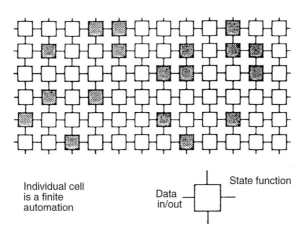

Individual cell is a finite automation

Data in/out

State function

Fig. 14 A lattice behaving as a two-state automaton. Reproduced by permission from *Introduction to molecular electronics*, M. Petty, M.R. Bryce, and D. Bloor, Edward Arnold, London, 1995, p. 357.

Conway's famous *Game of life*, a program available for most personal computers.

Having set up the lattice on the screen one defines an algorithm which switches the state of a given element when a given number of neighbouring elements are in the same state. Starting with a group of elements in the 'on' state (symbolizing for us the injection of a signal, for instance at the edge of the array) the pattern evolves from generation to generation, perhaps exiting on the other side of the array. Figure 15 shows an example. This idea is not new; it has been investigated theoretically by Hopfield and Wolfram in the USA, and in the UK by Barker. What has it got to do with molecular-based information processing? Well, the smallest kind of regular array of objects is a crystalline one formed from molecules, each of which might have a different charge, spin, vibration, conformation, protonation, or any other kind of state we might dream up. The beauty of molecular arrays

Fig. 15 Evolution of patterns in a two-state automation. Reproduced by permission from *Introduction to molecular electronics*, M. Petty, M.R. Bryce, and D. Bloor, Edward Arnold, London, 1995, p. 357.

is that, as we have seen, for example in Fig. 8, given a particular shape and charge distribution among the constituents, they assemble themselves, in contrast to the need for lithographic writing to create patterns on silicon. But now we are straying towards science fiction, and it is the moment to sum up.

Facts for the future

As a platform for thinking, and perhaps also action, towards an artificial information storage and processing regime based on molecular ingredients, let us consider a number of clear facts. First, since the dawn of computation, the individual information storage and processing entities found in computing machines have been inexorably getting smaller and smaller. Second, chemistry is good at making precisely defined and structured arrays of molecules, capable of existing in different electronic, magnetic or whatever kinds of distinguishable states. Third, methods do exist whereby single molecules can be addressed, although fourth, and more subtly, perhaps it is not necessary to do so. Does all that make it likely that we shall see molecular-based information processing in practise one day? Finally, of course, we do not know but, for myself, if I could be here in 50 years time to collect, I would bet on it. After all, our brains make a pretty good job of it already!

Acknowledgements

For lending materials and examples to illustrate this presentation, my warmest thanks go to Professors Richard Catlow and Olivier Kahn.

Bibliography

The following are some general references to molecular electronics and information processing:

M.C. Petty, M.R. Bryce, and D. Bloor (ed.), *Introduction to molecular electronics*, London, Edward Arnold, 1995.

P. Day, Future molecular electronics: towards a supramolecular information processor, *Chemistry in Britain*, 1990, **26**, 52–4.

P. Day, Room at the bottom, *Chemistry in Britain*, 1996, **32**, 29–31.

R.M. Metzger, P. Day, and G. Papavassilliou (ed.), *Low dimensional structures and molecular devices*, New York, Plenum Press, 1990.

PETER DAY

Born 1938 in Kent and educated at the local village primary school and nearby grammar school at Maidstone. An undergraduate at Wadham College, Oxford, of which he is now an Honorary Fellow. His doctoral research, carried out in Oxford and Geneva, initiated the modern day study of inorganic mixed valency compounds. From 1965 to 1988 he was successively Departmental Demonstrator, University Lecturer and An Hominem Professor of Solid State Chemistry at Oxford, and a Fellow of St John's College (Honorary Fellow 1996). Elected Fellow of the Royal Society in 1986; in 1988 he became Assistant Director and in 1989 Director of the Institut Laue-Langevin, the European high flux neutron scattering centre in Grenoble. In October 1991, he was appointed Director of The Royal Institution and Resident Professor of Chemistry, and Director of the Davy Faraday Research Laboratory, and in September 1994, he became Fullerian Professor of Chemistry. His present research centres on the synthesis and characterization of (mainly molecular) inorganic and metal-organic solids in the search for unusual magnetic and electron transport (including superconducting) properties.

'There or thereabouts'

ANDREW WALLARD

One of the books which fascinated many youngsters in the 1950s was Lancelot Hogben's *Man must measure*. In it, Hogben ranged freely and his message was that the ability to measure precisely was an important aspect of society and that it helped social, cultural, and industrial progress. He pointed out how important measurement was to trade and how it underpinned or helped develop many of the major inventions and innovations on which we have now come to depend, or has enabled companies to improve their manufacturing processes so as to make better products. One my themes will be the impact measurement technology has on almost every aspect of our lives.

Measurements' impact on daily life

As far as daily life is concerned, most people have rarely, if ever, given measurement more than a passing thought. Why should they? One of the reasons is because, one way or another, rather a lot of care and rather large sums of money are spent to ensure that daily measurements are reliable, routine, and nothing goes wrong. But by the same token a lot of things on which we rely implicitly have immense potential for 'going wrong' if the system doesn't work property.

For example:

1. When paying an electricity bill or filling up a motor car consumers instinctively rely on the meter reading or the pump gauges for the price they pay. Petrol pumps are accurate to about 0.3% which translate to 50p or so a year for the average motorist; just a bit less, incidentally, than each man, woman, and child pays to run the National Physical Laboratory (NPL).

2. One-quarter of a million people have cancer therapy each year. The dose needs to be measured to about ± 5% to make sure the cancer

cells and not the live tissue is damaged. Precise measurement is important and even the best standards laboratories are actually hard pressed to calibrate machines to better than 1.5%.

3. Aircraft altimeters need regular calibration by the world's airlines to the same specification.

4. Better quality and precision engineering have ensured that motor cars now have much longer service intervals.

Measurement, then, is all around us and we automatically rely on it as one of the 'givens' of today's society. To begin with, let us look back at the evolution of measurement before we set it into today's context.

Early measurement: use of the body

Nobody really knows much about the sort of early measurement which featured in the first few pages of Hogben though the two most important factors which made it important were construction and trade. The *Bible* is littered with references to various measures of length and weight (Noah's ark was 300 cubits long, 50 wide, and 30 high—*Genesis*) and we certainly know of sophisticated measurements which go back to between 3000 and 4000 BC. The Egyptian Pyramids were built about 2500 BC and required length measurements to about one-tenth of a 0.1%. Most early length measurements were based, of course, on the body with the cubit being the distance between fingertips and elbow. This use of the body to measure things is still reflected in some of today's language, for example, the hand, the pace, the mile (*mille passum* or 1000 paces) were all dimensional measures.

Weighing was important because of its link with buying, selling, and trading. Rather like parts of the body being used for length measures the ancient civilizations turned to other 'artefacts of nature' for other quantities. In the case of weight, the most common 'standard' was a seed. The carob seed was common in the Mediterranean and is now what we call the carat used to weigh gold and silver. Other popular, and in fact quite reproducible seeds, were grains of wheat or barley. (They were also used for length—three seeds laid end to end was 1 inch.) The most widely used heavier weights were based on the shekel—256 barley or wheat ears. On a practical scale there was a link with coinage and a penny-weight in the year 1200 was based on 32 grains of wheat, although 250 years later the pennyweight was only 12 grains because the amount of silver in it had been reduced. Other systems built on this standard, and 2560 pennyweights were the weight of 1 gallon of wine—so in actual fact barley corns could act as the basis of measures for length, weight, and volume.

The hallmark, though, of early measurement was that there was no single system and even great Empires like the Greeks and Romans failed to establish consistency. Of the two, the Greek civilization (700 BC to AD 400) had the greater scientific culture and although the Athenian or Attic standards dominated domestically, the ancient Greeks, because they were major traders, compared their own standards carefully with others and also kept copies of the alternative Persian and Phoenician systems in Athens. The Romans, on the other hand, (300 BC to AD 400), although they were superb engineers, had virtually no scientific traditions so they borrowed most of their weights and measures infrastructure from the Greeks. In the case of the countries they conquered they merely took the nearest local equivalent to their own standard measure, and called it by the name of their own unit. This led to immense confusion and there was, for example, no single 'Roman pound' in the sense that there *was* an 'Athenian' one, and no attempt to reflect the Greek practice of denoting their official standards with some sort of embossed seal.

Some semblance of order

Closer to home in Britain, there was a similarly rather chaotic situation until, somewhere about the time of the Norman Conquest, William I tried to establish some consistency on to a mix of regional, Saxon and Roman-based systems and ordered that the existing weights and measures had to be 'duly certified' or compared with each other. Even then, all was a bit arbitrary and it was about this time that William of Malmsbury recorded Henry I's declaration that the yard was the distance between the Royal nose and finger. We had to wait until the *Magna Carta* (1215) for the first government policy statement on metrology ('one measure to be established throughout the land') and real attempts at some degree of consistency. Like most government statements it was applied patchily but unique 'national' standards began to be kept and were authorized by the Crown. A hundred years later in 1340, Edward III issued a Statute to authorize the 'Treasurer' (or the then Chancellor of the Exchequer) to make one set of bronze standards for the gallon, bushel, and of certain weights and to make copies for regional use. Most Monarchs then re-issued or confirmed these 'national' standards usually by stamping them. Several examples remain, notably the 1497 Winchester Yard of Henry VII, made of bronze and later endorsed by Elizabeth I.

Better standard measures

Most measurements only needed to be done to a per cent or so and bronze and iron yards, and weights were good enough for most practical purposes. But the mid-eighteenth century brought scientific and technical advances which began to put standards on to a more sound footing.

Fire destroyed Parliamentary copies

Different metals also began to be used because the length of the popular bronze or brass bars were too susceptible to temperature. In fact, the last formal bronze standard was the 1824 'Primary Standard Yard' kept in the House of Commons. This yard was destroyed in 1834 together with the pound when the Palace of Westminster was burned down.

Imperial standards

So, as a result of the fire, and in one of Parliament's brief flashes of interest in metrology, it set up a committee to derive newer, better standards and to ensure continuity with the old length and mass standards. Three Fellows of the Royal Society (Reverend R. Sheepshanks, Mr F. Baily, and Professor W. Miller) worked until 1854 to produce scientifically respectable national standards.

1. The *Imperial Standard Yard*, a solid square gunmetal bar made from what came to be called 'Mr Baily's metal' (16 proportions of copper, 2.5 of tin, and one of zinc) and which was stiff and less sensitive to temperature than bronze (eighty-five copper, fifteen tin) or brass (67 copper, 33 nickel). The critical dimension was between fine lines in two gold plugs, when the temperature was 62°F and the atmosphere pressure 30 inches of mercury;

2. The *Avoirdupois Pound* of platinum.

3. The *Imperial Gallon* which was 10 Avoirdupois pounds of pure water at 62°F and 30 inches of mercury.

The job of looking after these standards was given to the Board of Trade but prudence—the hallmark of all good committees—came into play and several copies were made and deposited with Parliament, the Royal Greenwich Observatory, the Royal Mint, and the Royal Society. To some extent, these remain our 'national' measures and copies are in the wall of Trafalgar Square.

Evolution of the metric system

To understand something of the evolution of our current system of scientifically based standards, we have to look to our Gallic neighbours. As far back as 1742, French and British scientists compare the *pied* and *livre* of Paris with the British foot and pound. Not surprisingly, given the way national systems had evolved, they differed by about 7%—far in excess of the uncertainties each was thought to have. The scientific challenge was to find a reference, preferably drawn from nature so that it was not bound to any particular country—a challenge, in fact, we still work towards today.

Length

The first standard to be tackled was length. There were two choices: one a pendulum which had a period of 2 seconds; the other a standard of length based on the length of the meridian which joined the earth's North and South Poles. The latter won the day because scientists already knew enough about pendulums to realize that the pendulum period depended on the acceleration due to gravity and so would change over the world. The French therefore pressed ahead with a decree which established the metre as one ten-millionth of the length of the portion of the earth's meridian from Dunkirk to Barcelona. Other 'metric' units followed—the mass unit, for example, was the mass of a cubic decimeter of water at 4°C.

Having stated, or defined, the standard metre, the task of actually measuring it also fell to the French. This they did by precise and careful triangulation (the technique still used by surveyors), and produced a platinum standard metre as well as a kilogram in 1799.

International standards/BIPM

These matters rested until the Paris World Exhibition in 1867 when a group of scientists met to assess the state of metrology. As a result, an improved metre with an x-shaped section (made famous in school textbooks) and a new kilogram were made from the same block of platinum–iridium metal. These formed the basis of the world-wide 'metric' system established under the *Convention du Metrè* by 20 countries in 1875. The 'international' standards were entrusted to the care of the *Bureau*

International des Poids et Mesures—or BIPM—in Paris where they reside to this day. The UK signed the *Convention du Metrè* in 1884. This is the foundation of today's international measurement system based on the 7 units of the metric system, themselves used to build up all other 'derived' units like pressure, force, electrical resistance, density, etc.

Twentieth century standards—practical, portable measurements available to all

The twentieth century saw rapid advances in the world's capability to measure. It also enabled metrologists to move away from unique 'physical artefact' standards to transportable devices, which can give access to highly accurate measurements in the scientific laboratory or in industry. As an example of this, I will concentrate on five of the seven 'base' or fundamental units.

Twentieth-century metrology

The *Convention du Metrè* signified the acceptance by most of the industrialized countries that they had to have consistent weights and measures. Most, if not all, traced their national standards to those at the BIPM and periodically took them there for re-calibration and checking.

New industry

During the early part of the century two important things happened. First, there was a considerable rise in industrial activity, and a rush of new products (transportation, aviation, communication, domestic equipment), which placed increasing demands on measurement standards as precision engineering improved and people needed new standards for a whole variety of things (radioactivity, lighting). The National Standards laboratories had to find new ways of improving the accuracy or stability of top level standards so as to satisfy the needs of the new users. In fact, this century has seen a regular and systematic improvement in the performance of national standards—so much so that we have had to improve them twofold about every 10 years. That is why we have, as some say, to keep on adding more decimal points!

Length standards—why was the metre bar not good enough?

The second thing was that quantum physics gave us the potential to replace material standards by ones based on the fundamental laws of physics. The things which limited the performance of standards like the metre were the fact that they expanded if the temperature rose, the metal itself changed length as a result of metallurgical changes, and the precision of the optical microscope techniques used to compare bars limited the precision with which intercomparisons between primary and secondary standard bars could be carried out. In addition it was extremely hard to define the 'shape' of the finely engraved line which marked the reference point.

Quantum effects—the Bohr model

To understand why quantum effects are important we need to spend a few moments looking at simple atomic theory. In an atom electrons spin around a nucleus. When the atom gains energy in some way—for example, it is accelerated, heated, or put in an electric discharge—then the electrons jump to wider orbits and to higher energy levels. They return to their natural state by losing the energy, usually in the form of light. That is what happens in a lamp or a fluorescent light tube. What quantum physics did was to teach us that the energy levels to which the electrons could jump were fixed for a particular atom and so the differences between them were precise, were 'set' by the natural laws of physics and were the same everywhere in the world. Mathematically, we describe this by the $E=hf$ relationship where E is the energy difference between the electron energy levels, h is a constant called Planck's constant and f the reciprocal of the wavelength or colour of the emitted light. The actual wavelength is about half a millionth of a metre and so we have, in principle, a very small and 'natural' standard or ruler.

Light sources

These discoveries encouraged considerable work into the development of lamps as light sources and wavelength standards. In most lamps the spectrum of the light emitted is a series of discrete 'lines' rather than the continuous spectrum which is what makes 'white' light. The scientific problem which had to be tackled was the breadth of the 'spectral lines' from early lamps. The lines were broad because the atoms in the lamp were moving in all directions—the so-called 'Doppler' effect. Much scientific effort was therefore directed towards the study of the stability of the lines from certain lamps to find out what determined their width and to try and find ways of making the lines (and hence the wavelength, or the divisions of the 'ruler') as narrow as possible.

Interferometers

It was worth persevering with these painstaking researches because the optical technique called 'interferometry' made it possible to carry out

practical, precise length and distance measurements using the new narrower light sources. The principle of interferometry is simple if we think
of light as a wave. The energy associated with light varies like a wave as
it moves through space. If we arrange things so that two similar light
waves overlap each other then the 'wave nature'—the ups and downs—
are reinforced. If we shift the two waves by exactly half of the regular
wave pattern (half a wavelength) then they cancel each other out. To
create this effect the light waves are only shifted by about one-third of a
millionth of a metre. A combination of narrow light sources and interferometry therefore offered the opportunity to measure 'standard' metre
bars in terms of this much more stable 'quantum' reference. Eventually,
it was possible metaphorically to throw away the 1889 metre and
redefine it in 1960 as '1650 763.73 wavelengths in vacuum of the radiation corresponding to the transition between the levels $2p_{10}$ and $5d_5$ of
the *kr 86* atom' (Fig. 1).

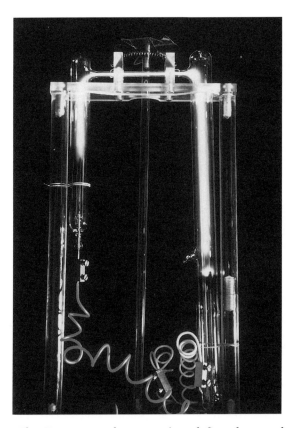

Fig. 1 The Krypton 86 lamp used to define the standard of
length between 1960 and 1983.

Krypton lamps

There was a price to pay for this greater accuracy and the 'universality' of the standard because the equipment became more and more complex. Although it was much more accurate and stable or reproducible than the metre bar, the krypton lamp was much more tricky to deal with and had to be operated under carefully controlled conditions. Fortunately, very soon after the international definition of the metre was changed the laser was invented and like so many advanced developments, was put to immediate use in precision metrology.

Iodine lasers—the current length standard

The advantages of light from a laser were that it could be, by its very nature, much narrower than the krypton line and that it was much brighter. Today, laser standards have superseded the krypton lamp.

The wavelength of light from a laser can be tuned by making changes to the distance between the mirrors. The challenge was to find some stable reference point within the tuning range and keep the laser tuned to it. The solution lay with the discovery that iodine vapour absorbed light from the helium–neon laser. Some careful spectroscopy and calculations showed that the centres of 10 iodine absorption lay within the turning range. However, just like the krypton lamp the iodine molecules were all moving round so even these absorption lines had a 'width'. The trick was to put a tube of iodine inside the laser, which meant that the laser light could be made to have a special interaction with the iodine atoms (saturated absorption) and could pick off only those molecules that moved at right angles to the laser beam. As a result the system could detect super-narrow lines. By keeping the laser adjusted so that its wavelength was at the centre of one of these absorption lines it was possible to create today's length standard reference—about a thousand times narrower than the krypton line (Fig. 2).

Electric standards—the Josephson volt

Until the late 1980s, the most reliable and stable voltage standards were Weston standard cells—small chemical batteries which, treated well, could be remarkably stable. But the search for better electrical standards was similar to that for better length ones. Metrologists wanted standards which did not vary with time or with the material from which they are made. Today's voltage standard achieves these objectives and is based on the Josephson effects in superconductivity. Devices, rather like a sandwich of two superconducting metals separated by a thin barrier of insulating material, have very interesting properties when kept at the

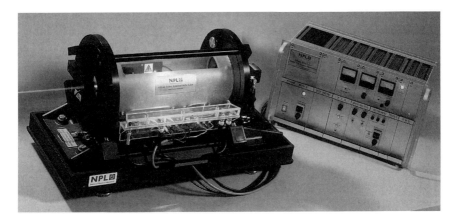

Fig. 2 The iodine stabilized laser used to realize today's definition of the metre.

temperature of liquid helium: −279°C. Although there is a physical barrier, two superconducting electrons pair up and can tunnel through between the two metals without developing a voltage between them. If a voltage is applied to the Josephson junction, the electron pairs tunnel to a state of different energy. Energy has, however, to be conserved and the electron pair has to lose some energy in the form of microwaves. The Josephson voltage standard reverses this process and shines microwave energy on to the junction. What actually happens is that the two effects mix with each other and result in very accurate quantized voltages which depend only on the applied frequency and various fundamental physical constants—e, the charge on the electron and h, Planck's constant. Because frequency is a very accurate quantity, we can now have an almost equally accurate voltage.

The voltage produced by one junction is extremely small, only about one ten-thousandth of a volt, so is itself hard to measure and, in practice, between 3000 and 20 000 individual junctions are put together so as to build up voltages of up to 10 V. The arrays are now made commercially and used in several industrial laboratories where high accuracy is needed, as well as in the majority of National Standards laboratories.

Time

One of the world's great clocks was Harrison's chronometer. In the mid-eighteenth century the government offered a prize for a marine clock. The target they set was for a loss of less than 2 minutes over a 6-week period. Harrison's clock lost only 15 seconds over 5 months. Today's time standard is, however, a caesium atomic clock. It works rather like

Fig. 3 Atomic clocks used at NPL to set the UK's time scale.

the laser length standard but we actually use microwaves rather than light to make the caesium atom jump between two of its energy levels. These clocks are the most accurate of all today's measurement standards—an uncertainty of the equivalent of only 1 second in 300 000 years. The beauty of this standard is that its easily made available to anybody with a radio receiver through what is called the 'Greenwich time signal' but which actually comes from NPL's clocks. (Fig. 3)

Temperature standards—the kelvin

Temperature is another of the 'base units'. Until 25 or so years ago most scientists thought 100°C was the boiling point of water and 0°C the freezing point. But as in other cases, better experiments in the science of thermodynamics started to challenge these 'practical' measurements and as a result the boiling point of steam turned out to be only 99.97°C! The international temperature scale had to be redefined because physicists are happiest with a sound thermodynamics scale even though to do 'real' measurements there are set fixed points.

The most common fixed point is known as a triple point. At a triple point a solid, liquid and a gas can coexist—a unique thermodynamic situation, which can be described by theories and made consistent with all the other measurements and standards. When a carefully prepared pure reference mixture is cooled the temperature drops with time until it reaches a stable 'plateau' when all the three phases exist. The international standard for the kelvin is defined using a triple point cell of highly pure water in an evacuated phial and is 1/273.16 of the thermodynamic temperature of pure water at its triple point. (Fig. 4)

For higher or lower temperatures, the international temperature scale uses several other triple and freezing or meeting points of the various

Fig. 4 Cell used to define the Kelvin at the triple point of water.

metals or mixtures used to set fixed points for very low to very high temperatures.

Mass standards—the kilogram

Next—and definitely *not* an atomic standard—is the kilogram. Quite remarkably, it is still based on its nineteenth century predecessor—a small cylinder of platinum iridium. There is a curious mystique about the kilogram. *The* kilogram is in Paris at the BIPM and has been since 1889. Each country has a copy which has been weighed carefully against it (Fig. 5). All are checked—three times in the last 120 years—to see how they drift around. Usually kilograms gain weight by absorbing gases or pollutants from the air—they are kept in air rather than in a vacuum because once a polluting layer has built up on the metal surface, it remains extremely stable.

Metrologists can compare kilograms to one-thousandth of a microgram— good enough for industrial and commercial measurement but a tiresome and painstaking measurement. There are, therefore, several attempts to relate the kilogram to the more accurate electrical standards, which look like promising replacements for the kilogram within the next 10 years.

Summary

We have concentrated—for practical reasons—on five of the seven international standards. The most accurate is the atomic clock closely

Fig. 5 The UK's copy of the international kilogram.

followed by the electrical and laser length standards. In nearly all cases industrial and scientific demands for improvements are increasing and so in some areas metrologists are already on the look out for new or improved ways of measuring.

Applications

I began by claiming that accurate measurement was at the heart of society and its industrial and commercial life. A few final examples can help illustrate just a few of today's measurement challenges.

Force measurement relates directly to the kilogram and metre (weight per unit area) and is one of the areas where industry is pressing national standards laboratories most vigorously for improved accuracy and range of measurement. Figure 6 shows NPL's 1.2 MN force machine (120 tonnes). The weights used to produce the forces are directly traceable to the kilogram and each is known to 3 p.p.m. In building it they had to be

Fig. 6 NPL's 1.2 Mn force machine being used to calibrate a load cell.

aligned to 1 mm and 'little g' (the acceleration due to gravity) measured over its 30 m height. The accuracy is about 10 p.p.m. and is needed:

(1) for measuring jet engine thrust—big engines have about 30 tonne thrusts, which have to be measured to half a tenth of a per cent so as to conform with customers' specifications;

(2) for weighing oil rigs, such as the Conoco plant in the North Sea, which has to be weighed so that the correct forces can be applied to keep it tethered to the sea bed; and

(3) to calibrate thrust transducers incorporated in 'Thrust 2' the car currently attempting to set a new world land speed record.

Length measurement is increasingly being applied to enable others to make precise measurements in industry and science. Figure 7 shows recent results from a Cambridge group which employed up to date interferometric technologies to make for the first time milli-arc second resolution measurements of the rotating double star, Capella.

Medical applications now demand carefully controlled ball/socket clearance of 10 μm in artificial hip joints.

In *time* measurement satellite-based clocks can be used to locate a position on the earth to about 3 m through the 'global positioning system' (GPS), which is also used to help control the Conoco oil rig

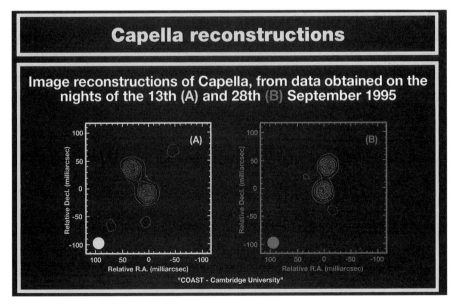

Fig. 7 Image reconstruction of the double star 'Capella' showing the rotation of the two stars with milli-arc second resolution.

discussed in the second point on force measurement. Advances in technology mean that it is now possible to buy hand-held GPS receivers which show latitude, longitude, and height.

Advances in satellite-based time measurement and transfer will mean that the twenty-first century aeroplanes will land and navigate by improved GPSs.

Conclusions

Measurement is, indeed, all around us and is subject to ever-increasing demands for improved accuracy. This trend is likely to continue and we expect to be challenged by new applications of measurement in a variety of different fields.

Readers of this record may like to know that the lecturer, knowing the importance of precise timing at the Friday Evening Discourses, presented the Royal Institution with a radio-controlled clock so that NPL's 'atomic' time signal could be used to calibrate the lecture theatre's clocks.

ANDREW WALLARD

Born 1945 and educated at the University of St Andrews. He joined the National Physical Laboratory in 1968 to work on interferometry and the use of frequency stabilized lasers as standards of length. 1978–1990 he was employed in a number of Whitehall positions including the Central Policy Review Staff and policy support groups for Ministers at the Department of Trade and Industry (DTI). He was responsible for several DTI research programmes in electronics and optoelectronics and for the European Union ESPRIT programme which complemented the UK's 'Alvey' IT programme. In 1990 he returned to NPL as Deputy Director, a post which he continues to hold following the privatization of NPL in 1995.

Pondering on Pisa

JOHN BURLAND

Introduction

In 1989 the civic tower of Pavia collapsed without warning, killing four people. The Italian Minister of Public Buildings and Works appointed a commission to advise on the stability of the Pisa tower. The commission recommended closure of the tower to the general public and this was instituted at the beginning of 1990. There was an immediate outcry by the Mayor and citizens of Pisa who foresaw the damage that the closure would inflict on the economy of Pisa, heavily dependent on tourism as it is. In March 1990 the Prime Minister of Italy set up a new commission, under the chairmanship of Professor Michele Jamiolkowski, to develop and implement measures for stabilizing the tower. It is the fifteenth commission this century and its membership covers a number of disciplines including structural and geotechnical engineering, architecture, architectural history, archaeology, and restoration.

It is instructive to imagine a tower, founded on jelly and slowly inclining to the point at which it is about to fall over. Any support would also have to rest on the jelly. Worse still, the masonry composing the tower is so fragile that it could explode at any time. This is a reasonable description of the state of the Leaning Tower of Pisa, and helps to explain why stabilizing it represents the ultimate civil engineering challenge.

Details of the tower and ground profile

Figure 1 shows a cross-section through the tower. It is nearly 60 m high and the foundations are 19.6 m in diameter. The weight of the tower is 14 500 t. At present the foundations are inclined due south at about 5.5° to the horizontal. The average inclination of the axis of the tower is somewhat less due to its slight curvature as will be discussed later. The seventh cornice overhangs the first cornice by about 4.5 m.

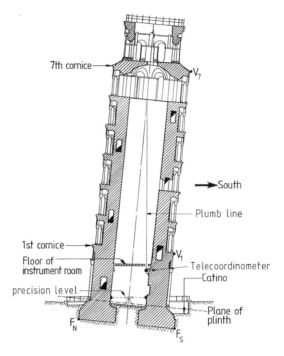

Fig. 1 Cross-section through tower.

Construction is in the form of a hollow cylinder. The inner and outer surfaces are faced with marble and the annulus between these facings is filled with rubble and mortar within which extensive voids have been found. The spiral staircase winds up within the annulus.

Figure 2 shows the ground profile underlying the tower. It consists of three distinct layers. Layer A is about 10 m thick and primarily consists of estuarine deposits laid down under tidal conditions. As a consequence the soil types consist of rather variable sandy and clayey silts. At the bottom of layer A is a 2 m thick, medium dense, fine sand layer (the upper sand). Based on sample descriptions and cone tests, the material to the south of the tower appears to be more clayey than to the north and the sand layer is locally much thinner. Therefore, to the south layer A could be expected to be slightly more compressible than to the north.

Layer B consists of soft, sensitive, normally consolidated marine clay, which extends to a depth of about 40 m. The upper clay, known as the Pancone Clay, is very sensitive to disturbance which causes it to lose strength. The lower clay is separated from the Pancone Clay by a sand layer (the intermediate sand) overlain by a layer of stiffer clay (the intermediate clay). The Pancone Clay is laterally very uniform in the vicinity

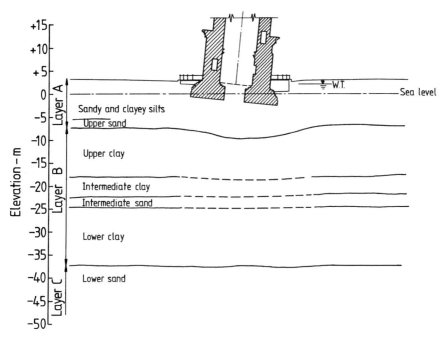

Fig. 2 Soil profile beneath tower.

of the tower. Layer C is a dense sand which extends to considerable depth (the lower sand).

The water table in horizon A is between 1 and 2 m below ground surface. Pumping from the lower sand has resulted in downward seepage from layer A with a vertical pore pressure distribution through layer B, which is slightly below hydrostatic.

The many borings beneath and around the tower show that the surface of the Pancone Clay is dished beneath the tower from which it can be deduced that the average settlement is between 2.5 and 3 m.

History of construction

The tower is a campanile for the cathedral, construction of which began in the latter half of the eleventh century. Work on the tower began on 9 August 1173 (by the modern calendar). By about 1178 construction had progressed to about one-quarter of the way up the fourth storey when work stopped. The reason for the stoppage is not known but had it continued much further the foundations would have experienced a bearing capacity failure within the Pancone Clay. The work recommenced in about 1272, after a pause of nearly 100 years, by which time the strength of the clay had increased due to consolidation under the weight of the tower. By about 1278 construction had reached the seventh cornice when work again stopped—possibly due to military action. Once again there can be no doubt that, had work continued, the tower would have fallen over. In about 1360 work on the bell chamber was commenced and was completed in about 1370—nearly 200 years after commencement of the work.

It is known that the tower must have been tilting to the south when work on the bell chamber was commenced as it is noticeably more vertical than the remainder of the tower. Indeed, on the north side there are four steps from the seventh cornice up to the floor of the bell chamber while on the south side there are six steps. Another important historical detail is that in 1838 the architect Alessandro Della Gherardesca excavated a walk-way around the foundations. This is known as the catino and its purpose was to expose the column plinths and foundation steps for all to see as was originally intended. This activity resulted in an inrush of water on the south side, as here the excavation is below the water table, and there is evidence to suggest that the inclination of the tower increased by as much as half a degree. As is described later, Gherardesca left us an unpleasant surprise.

History of tilting

One of the first actions of the commission was to undertake the development of a computer model of the tower and the underlying ground that could be used to assess the effectiveness of various possible remedial measures. Calibration of such a model is essential and the only means of doing this is to attempt to simulate the history of tilting of the tower during and subsequent to its construction. Hence it became apparent very early on that we needed to learn as much as possible about the history of the tilt of the tower. In the absence of any documentary evidence all the clues to the history of tilt lie in the adjustments made to the masonry layers during construction and in the shape of the axis of the tower.

Over the years a number of measurements of the dimensions of the tower have been made and many of them are conflicting. The Polvani Commission measured the thickness of each of the masonry layers and its variation around the tower.[1] This information has proved extremely valuable in unravelling the history of tilt.

Figure 3 shows the shape of the axis of the tower deduced from the measured relative inclinations of the masonry layers assuming that con-

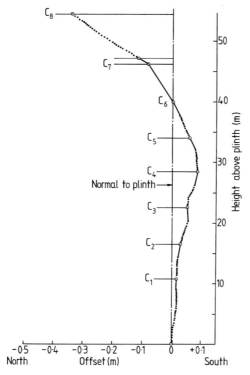

Fig. 3 Shape of axis of tower deduced from relative inclinations of masonry layers.

struction proceeded perpendicular to each masonry layer. This shape compares favourably with other independent measurements at a few locations up the tower. It can be seen that the axis is curved. For years the tower has been unkindly referred to as having a banana shape. I prefer to call it a question mark (?) so as to reflect the enigma of the tower.

Some important observations can be made from the measurements on the masonry layers. For most of the storeys, construction took place using parallel sided blocks of masonry. With one or two notable exceptions adjustments only took place close to each floor using tapered blocks. The most important exception can be seen in Fig. 3 where there is an obvious kink one-quarter the way up the fourth storey. It will be recalled that the construction remained at this level for about 100 years. Evidently, the tower was tilting significantly when work recommenced and the masons made adjustments to correct it.

We see that the history of the tilting of the tower is tantalizingly frozen into the masonry layers. If only we knew the rules that the masons followed in adjusting for the tilt we would be able to unravel the history.

We have to put ourselves in the place of a mason or architect in the twelfth or thirteenth century and ask ourselves: 'What is the most practical thing to do when you arrive at a given floor and find that the tower is out-of-plumb?' A widely accepted hypothesis is that the masons would always try to keep the masonry layers horizontal and the Polvani Commission adopted this. Although this seems reasonable for a low aspect ratio building like a cathedral, it does not make sense for a tower as it would tend to perpetuate the overall out-of-plumb. After a few trials, a child building a tower of bricks on a carpet will soon learn to compensate for any tilt by attempting to place successive bricks over the centre of the base of the tower, i.e. by bringing the centre of the tower back vertically over the centre of the foundations (or possibly even further, away from the direction of tilt). Therefore, an alternative hypothesis is one in which the masons aimed to bring the centre line of the tower back, vertically over the centre of the foundations at the end of each storey. The architectural historians on the commission are satisfied that the masons would have had the technology to make such an adjustment, particularly as the stones for each storey were carved and assembled on the ground prior to hoisting into position.

Figure 4 shows the re-constructed history of inclination of the foundations of the tower using the alternative hypothesis.[2] In this figure the

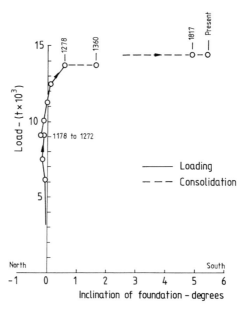

Fig. 4 Deduced history of inclination of tower during and subsequent to construction.

weight of the tower at any time is plotted against the deduced inclination. It can be seen that initially the tower inclined slightly to the north
amounting to about 0.2° in 1272 when construction recommenced. As
construction proceeded the tower began to move towards the south at an
increasing rate. In 1278, when construction had reached the seventh
cornice, the tilt was about 0.6°. During the 90-year pause, the tilt
increased to about 1.6°. After the completion of the bell chamber in
about 1370 the inclination of the tower increased dramatically. The
point dated 1817 is based on measurements made by two British archi-
tects Cressy and Taylor using a plumb line. A further measurement was
made by the Frenchman Ruhault de Fleury in 1859, which showed that
the excavation of the catino by Gherardesca in 1838 caused a significant
increase of inclination. The history of tilting depicted in Fig. 4 has been
used to calibrate numerical and physical models of the tower and underlying ground.

Computer modelling of the movements of the tower

The analysis was carried out using a suite of finite element geotechnical
computer programs developed at Imperial College and known as ICFEP.[3]
The constitutive model is based on critical state concepts[4] and is non-
linear elastic work-hardening plastic. Fully coupled consolidation is
incorporated so that time effects due to the drainage of pore water out of
or into the soil skeleton are included.

It must be emphasized that the prime objective of the analysis was to
develop an understanding of the mechanisms controlling the behaviour
of the tower.[5] Accordingly, a plane strain approach was used for much
of the work and only later was three-dimensional analysis used to
explore certain detailed features.

The layers of the finite element mesh matched the soil sublayering
that had been established from numerous extensive soil exploration
studies. Figure 5 shows the mesh in the immediate vicinity of the foundation. In layer B (see Fig. 2) the soil is assumed to be laterally homoge-
neous. However, a tapered layer of slightly more compressible material
was incorporated into the mesh for layer A1 as shown by the shaded
elements in Fig. 5. This slightly more compressible region represents the
more clayey material found beneath the south side of the foundation as
discussed in the second section. In applied mechanics terms the insertion of this slightly more compressible tapered layer may be considered
to act as an 'imperfection'. The overturning moment generated by the
lateral movement of the centre of gravity of the tower was incorporated

into the model as a function of the inclination of the foundation as shown in Fig. 5.

The analysis was carried out in a series of time increments in which the loads were applied to the foundation to simulate the construction history of the tower. The excavation of the catino in 1838 was also simulated in the analysis. Calibration of the model was carried out by adjusting the relationship between the overturning moment generated by the centre of gravity and the inclination of the foundation. A number of runs were carried out with successive adjustments being made until good agreement was obtained between the actual and the predicted present-day value of the inclination.

Figure 6 shows a graph of the predicted changes in inclination of the tower against time, compared with the deduced historical values. It is important to appreciate that the only point that has been predetermined in the analysis is the present-day value. The model does not simulate the initial small rotation of the tower to the north. However, from about 1272 onwards there is remarkable agreement between the model and the historical inclinations. Note that it is only when the bell chamber was added in 1360 that the inclination increases dramatically. Also of considerable interest is the excavation of the catino in 1838 which results in a predicted rotation of about 0.75°. It should be noted that the final imposed inclination of the model tower is 5.44°, which is slightly less than the present-day value of 5.5°. It was found that any further increase in the final inclination of the model tower resulted in instability—a clear indication that the tower is very close to falling over.

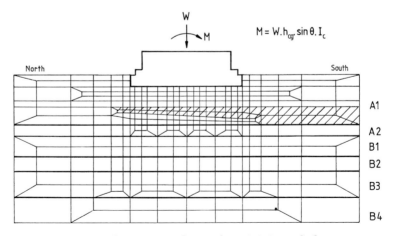

Fig. 5 Finite element mesh in the vicinity of the tower foundation.

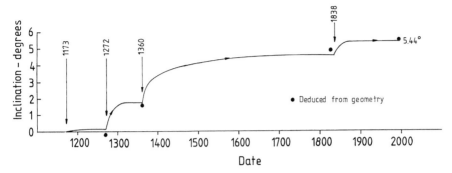

Fig. 6 Relationship between time and inclination for the computer simulation of the history of the Pisa Tower.

Burland and Potts[5] concluded from a careful study of the computer model that the impending instability of the tower foundation is not due to a shear failure of the ground but can be attributed to the high compressibility of the Pancone Clay. This phenomenon was called 'leaning instability' by the late Edmund Hambly[6] who used it to explain the lean of the Pisa tower. No matter how carefully the structure is built, once it reaches a critical height the smallest perturbation will induce leaning instability. As pointed out by Hambly: '... leaning instability is not due to lack of strength of the ground but is due to insufficient stiffness, i.e. too much settlement under load'. Children building brick towers on a soft carpet will be familiar with this phenomenon!

In summary, the finite element model gives remarkable agreement with the deduced historical behaviour of the tower. It is important to emphasize that the predicted history of foundation inclinations and overturning moments were self-generated and were not imposed externally in a predetermined way. The only quantity that was used to calibrate the model was the present-day inclination. The analysis has demonstrated that the lean of the tower results from the phenomenon of settlement instability due to the high compressibility of the Pancone Clay. The role of the layer of slightly increased compressibility beneath the south side of the foundations is to act as an 'imperfection'. Its principal effect is to determine the direction of lean rather than its magnitude. The main limitation of the model is that it is a plane strain one rather than fully three-dimensional. Also, the constitutive model does not deal with creep so that no attempt has been made to model the small time-dependent rotations that have been taking place this century and which are described in the next section. Nevertheless, the model provides important insights into the basic mechanisms of behaviour and has proved valuable in assessing the effectiveness of various proposed stabilization measures. Its role in evaluating the effectiveness of the temporary counterweight solution is described later.

Observed behaviour of the tower this century

Change of inclination

For most of this century the inclination of the tower has been increasing. The study of these movements has been important in developing an understanding of the behaviour of the tower and has profoundly influenced the decisions taken by the commission. It is important to appreciate that the magnitudes of the movements are about three orders of magnitude less than the movements that occurred during construction. Thus changes in inclination are measured in arc seconds rather than degrees (one arc second equals 1/3600th of a degree).

Figure 7 is a plan view of the *Piazza dei Miracoli* showing the location of the baptistry, cathedral, and tower. Since 1911 the inclination of the tower has been measured regularly by means of a theodolite. The instrument is located at the station marked E and the angles between station D and the first cornice (V1 in Fig. 1) and between station D and the seventh cornice (V7) are measured. The difference between these two angles is used to calculate the vertical offset between the seventh and first cornice and hence the inclination of the tower.

In 1928 four levelling stations were placed around the plinth level of the tower and were referred to a bench mark on the baptistry. Readings

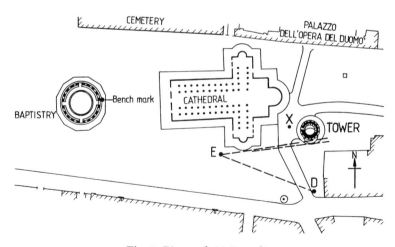

Fig. 7 Piazza dei Miracoli.

were taken in 1928 and 1929 but not again until 1965. In 1965, 15 level-ling points were installed around the tower at plinth level and about 70 surveying monuments were located around the piazza.

In 1934 a plumb line was installed in the tower, suspended from the sixth floor and observed in an instrument room whose location is shown in Fig. 1. The instrument was designed by the engineers Girometti and Bonechi and is known as the GB pendulum. Also in 1934 a 4.5 m long spirit level was installed within the instrument room. The instrument rests on brackets embedded in the masonry and can be used to measure both the north–south and the east–west inclination of the tower. The instrument was designed by the officials of the Genio Civile di Pisa and is known as the GC level.

Figure 8 shows the change of inclination with time since 1911. From 1934 to 1969 the GC level was read regularly once or twice a year except during the Second World War. For some reason readings with the GC level ceased in 1969, but fortunately precision levelling on the 15 points around the tower began in 1965 and continued regularly until 1985. In 1990 Professor Carlo Viggiani and I read the GC level again and found that the inclination agreed to within a few seconds of arc with that derived from the precision levelling around the plinth.

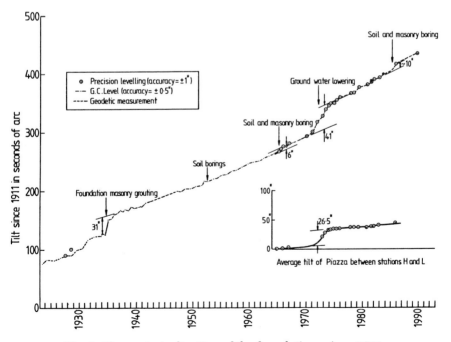

Fig. 8 Change in inclination of the foundations since 1911.

It can be seen from Fig. 8 that the inclination–time relationship for the tower is not a smooth curve but contains some significant 'events'. In 1934, Girometti drilled 361 holes into the foundation masonry and injected about 80 t of grout with a view to strengthening the masonry. This activity caused a sudden increase in tilt of 31 arc seconds. Like Gherardesca, Girometti also left us an unpleasant surprise as is described later. In 1966 some soil and masonry drilling took place and caused a small but distinct increase of tilt of about 6 arc seconds. Again, in 1985 an increase in tilt of 10 arc seconds. resulted from masonry boring through the foundations. In the late 1960s and early 1970s pumping from the lower sands caused subsidence and tilting towards the south-west of the piazza. This induced a tilt of the tower of about 41 arc seconds. When pumping was reduced the tilting of the tower reduced to its previous rate. It is clear from these events that the inclination of the tower is very sensitive to even the smallest ground disturbance. Hence any remedial measures should involve a minimum of such disturbance. The rate of inclination of the tower in 1990 was about 6 seconds of arc per annum or about 1.5 mm at the top of the tower.

The motion of the tower foundation

Previously, studies have concentrated on the changes of inclination of the tower. Little attention has been devoted to the complete motion of the foundations relative to the surrounding ground. The theodolite and precision levelling measurements help to clarify this. It will be recalled from Fig. 7 that angles were measured relative to the line ED. Hence it is possible to deduce the horizontal displacements of the tower relative to point D. Figure 9 shows a plot of the horizontal displacement of point

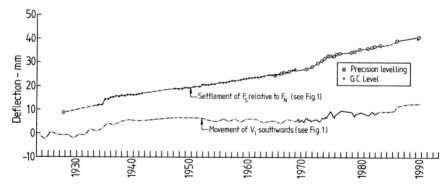

Fig. 9 Horizontal displacement of V1 on the first cornice since 1911.

V_1 on the first cornice relative to point D since 1911. Also shown, for comparison, is the relative vertical displacements between the north and south sides of the foundation (points F_N and F_S). It can be seen that up to 1934 the horizontal movement of V_1 was very small. Between 1935 and 1938, following the work of Girometti, point V_1 moved southwards by about 5 mm. No further horizontal movement took place until about 1973 when a further southward movement of about 3–4 mm took place as a result of the groundwater lowering. A further small horizontal movement appears to have taken place in about 1985 as a result of masonry drilling at that time. These observations reveal the surprising fact that during steady-state creep-rotation point V_1 on the first cornice does not move horizontally. Horizontal movements to the south only take place when disturbance to the underlying ground takes place.

Study of the precision levelling results shows that between 1928 and 1965 the centre of the foundations at plinth level rose by 0.3 mm relative to the baptistry—a negligible amount. Between 1965 and 1986 the relative vertical displacement between the centre of the plinth and a point a few metres away from the tower was again negligible. Thus, not only does point V_1 not move horizontally during steady-state creep, but also negligible average settlement of the foundations has taken place relative to the surrounding ground.

The observations described above can be used to define the rigid-body motion of the tower during steady-state creep-rotation as shown in Fig. 10. It can be seen that the tower must be rotating about a point approximately located level with point V_1 and vertically above the centre of the foundation. The direction of motion of points F_N and F_S are

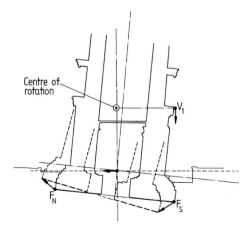

Fig. 10 Motion of the tower during steady-state creep-rotation.

shown by vectors and it is clear that the foundations are moving north-wards with F_N rising and F_S sinking.

Conclusions from the observed motion
of the tower foundations

The discovery that the motion of the tower is as shown in Fig. 10 has turned out to be a most important finding in a number of respects. Previously, it had been believed that the foundations were undergoing creep settlements with the south side settling more rapidly than the north. However, the observation that the north side had been steadily rising led to the suggestion that the application of load to the foundation masonry on the north side could be beneficial in reducing the overturning moment.[8]

The form of foundation motion depicted in Fig. 10 leads to the very important conclusion that the seat of the continuing long-term tilting of the tower lies in horizon A and not within the underlying Pancone Clay as had been widely assumed in the past. It can therefore be concluded that this stratum must have undergone a considerable period of ageing since last experiencing significant deformation. Thus, in developing the computer model, it is reasonable to assume that the clay has an increased resistance to yield subsequent to the excavation of the catino in 1838. This conclusion has proved of great importance in the success-ful analysis of the effects of applying the lead counterweight.[5]

The continuing foundation movements tend to be seasonal. Between February and August each year little change in the north–south inclina-tion takes place. In late August or early September the tower starts to move southward and this continues through till December or January amounting to an average of about 6 arc seconds. In the light of the observed motion of the tower foundations, the most likely cause of these seasonal movements is thought to be the sharp rises in groundwater level that have been measured in horizon A resulting from seasonal heavy rainstorms in the period September to December each year. Thus, continuing rotation of the foundations might be substantially reduced by controlling the water table in horizon A in the vicinity of the tower.

Temporary stabilization of the tower

There are two distinct problems that threaten the stability of the tower. The most immediate one is the strength of the masonry. It can be seen from the cross-section in Fig. 1 that at first floor level there is a change in cross-section of the walls. This gives rise to stress concentrations at the south side. In addition to this, the spiral staircase can be seen to pass through the middle of this change in cross-section giving rise to a significant magnification in the stresses. The marble cladding in this location shows signs of cracking. It is almost impossible to assess accurately the margin of safety against failure of the masonry, but the consequences of failure would be catastrophic. The second problem is the stability of the foundations against overturning.

The approach of the commission to stabilization of the tower has been a two-stage one. The first stage has been to secure an increase in the margin of safety against both modes of failure as quickly as possible by means of temporary measures. Having achieved this, the second stage is to develop permanent solutions recognizing that these would require time to carry out the necessary investigations and trials. Significant progress has been made with the first stage. It is a prerequisite of restoration work that temporary works should be non-destructive, reversible, and capable of being applied incrementally in a controlled manner.

Temporary stabilization of the masonry

The masonry problem has been tackled by binding lightly pre-stressed plastic covered steel tendons around the tower at the first cornice and at intervals up the second storey as shown in Fig. 11. The work was carried out in the summer of 1992 and was effective in closing some of the

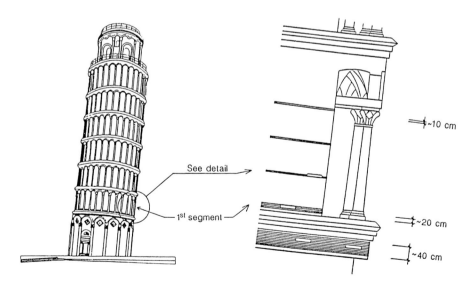

Fig. 11 Temporary stabilization of the masonry with light circumferential pre-stressing.

cracks and in reducing the risk of a buckling failure of the marble cladding. The visual impact has proved to be negligible.

Temporary stabilization of the foundations

As mentioned previously, the observation that the northern side of the foundation had been steadily rising for most of this century led to the suggestion that application of load to the foundation masonry on the north side could be beneficial in reducing the overturning moment. Clearly, such a solution would not have been considered if it had not been recognized that leaning instability rather than bearing capacity failure was controlling the behaviour of the tower or if the north side of the foundation had been settling.

Before implementing such a solution it was obviously essential that a detailed analysis should be carried out. The purpose of such an analysis was twofold: (i) to ensure that the proposal was safe and did not lead to any undesirable effects, and (ii) to provide a best estimate of the response against which to judge the observed response of the tower as the load was being applied. A detailed description of the analysis is given by Burland and Potts[5] who found that a satisfactory result was only forthcoming if the effects of ageing of the underlying Pancone Clay was incorporated in the computer model. The justification for such ageing lay in the observed motion of the foundations depicted in Fig. 10 as described in the previous section. The computer analysis indicated that it was safe to apply up to a maximum of 1400 t load to the north side of the foundation masonry. Above that load there was a risk that the underlying Pancone Clay would begin to yield resulting in a southward rotation of the tower and excessive settlement of the foundations.

Accordingly, a design was developed by Professors Leonhardt and Macchi for the application of a north counterweight and the details of construction are shown in Fig. 12. It consists of a temporary pre-stressed concrete ring cast around the base of the tower at plinth level. This ring acts as a base for supporting specially cast lead ingots, which were placed one at a time at suitable time intervals. The movements experienced by the tower are measured with a highly redundant monitoring system consisting of the following: (i) precision inclinometers and levellometers installed on the wall of the ground floor room; (ii) high precision levelling of eight survey stations mounted on the wall of the above room; and (iii) external high precision levelling of 15 bench marks located around the tower plinth and 24 bench marks located along north–south and east–west lines centred on the tower. All the levels are related to a deep datum installed in the *Piazza dei Miracoli* by the commission.

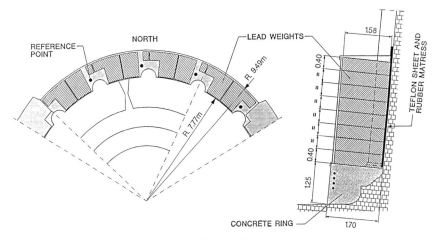

Fig. 12 Details of the north counterweight.

Observed response

Burland *et al.*[7] describe the response of the tower to the application of the counterweight. Construction of the concrete ring commenced on 3 May 1993 and the first lead ingot was placed on 14 July 1993. The load was applied in four phases with a pause between each phase to give time to observe the response of the tower. The final phase was split in two either side of the Christmas break. The last ingot was placed on 20 January 1994.

Figure 13 shows the change of inclination of the tower towards the north during the application of the lead ingots as measured by the internal high precision levelling and the inclinometer placed in the north–south plane. The agreement between the two independent monitoring systems is excellent. (Note that Fig. 15 does not include the inclination induced by the weight of the concrete ring, which amounted to about 4 arc seconds.) It can be seen that the amount of creep between the phases of load is small. However, subsequent to completion of loading, time-dependent northward inclination has continued. On 20 February 1994 (1 month after completion of loading) the northward inclination was 33 arc seconds. By the end of July 1994 it had increased to 48 arc seconds giving a total of 52 arc seconds including the effect of the concrete ring. On 21 February 1994 the average settlement of the tower relative to the surrounding ground was about 2.5 mm.

Comparison between predictions and observations

Figure 16 shows a comparison of the predictions from the computer model and measurements of: (i) the changes in inclination, and (ii) the

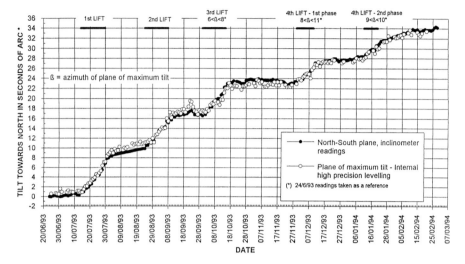

Fig. 13 Observed change of inclination of the tower during application of the counterweight.

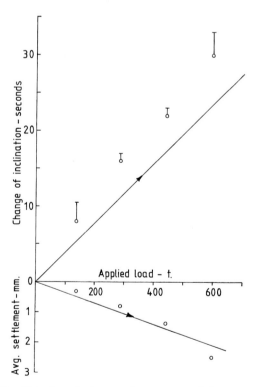

Fig. 14 Predicted and observed response of the tower to the counterweight.

average settlements of the tower relative to the surrounding ground during the application of the lead ingots. The points in the upper part of Fig. 16 represent the measured rotations at the end of each phase of loading and the vertical lines extending from them show the amount of creep movement between each phase. For the final phase the creep after 1 month is shown. It can be seen that the predictions of the computer model give changes in inclination, which are about 80% of the measured values. However, the predicted settlements are in excellent agreement with the measurements.

It is perhaps worth emphasizing that the purpose of the computer model was to clarify some of the basic mechanisms of behaviour and it was calibrated against inclinations measured in degrees. The use of the model in studying the effects of the counterweight was to check that undesirable and unexpected responses of the tower did not occur. In this respect the model has proved to be very useful. It has led to a considera-tion of the effects of ageing and it has drawn attention to the importance of limiting the magnitude of the load so as to avoid yield in the under-lying Pancone Clay. It is perhaps expecting too much of the model for it to make accurate quantitative predictions of movements that are three orders of magnitude less than those against which it was calibrated and the fact that it has done as well as it has is remarkable. The observed movements due to the application of the counterweight have been used to refine the model further.

Permanent stabilization

For bureaucratic and financial reasons work on the temporary stabiliza-tion of the tower has taken longer than had been hoped. In parallel with these operations the commission has been exploring a variety of approaches to permanently stabilizing the tower. The fragility of the masonry, the sensitivity of the underlying clay and the very marginal stability of the foundations has already been referred to. Because of these severe restraints, any measures involving the application of concentrated loads to the masonry or underpinning operations beneath the south side of the foundation have been ruled-out. Moreover, aesthetic and conserv-ation considerations require that the visible impact of any stabilizing measures should be kept to an absolute minimum.

The commission has decided to give priority to so-called 'very soft' solutions aimed at reducing the inclination of the tower by up to half a degree by means of induced subsidence beneath the north side of the foundation without touching the structure of the tower. Such an approach allows the simultaneous reduction of both the foundation instability and the masonry overstressing with a minimum of work on the tower fabric itself.

Some of the key requirements of stabilization by reducing inclination are as follows.

1. The method must be capable of application incrementally in very small steps.

2. The method should permit the tower to be 'steered'.

3. It must produce a rapid response from the tower so that its effects can be monitored and controlled.

4. Settlement at the south side must not be more than 0.25 of the settlement of the north side. This restriction is required to minimize damage to the catino and disturbance to the very highly stressed soil beneath the south side.

5. There must be no risk of disturbance to the underlying Pancone Clay, which is highly sensitive and upon whose stiffness, due to ageing, the stability of the tower depends.

6. The method should not be critically dependent on assumed detailed ground conditions.

7. The impact of possible archaeological remains beneath the tower must be taken into account.

8. Before the method is implemented it must have been clearly demonstrated by means of calculation, modelling, and large-scale trials that the probability of success is very high indeed.

9. It must be demonstrated that there is no risk of an adverse response of the tower.

10. Any preliminary works associated with the method must have no risk of impact on the tower.

11. Methods which require costly civil engineering works prior to carrying out the stabilization work are extremely undesirable for a number of reasons.

After careful consideration of a number of possible approaches the commission chose to study three in detail.

1. The construction of a ground pressing slab to the north of the tower, which is coupled to a post-tensioned concrete ring constructed around the periphery of the foundations.

2. Consolidation of the Pancone Clay by means of carefully devised electro-osmosis.

3. The technique of soil extraction as postulated by Terracina[9] for Pisa and widely used in Mexico City to reduce the differential settlements of a number of buildings due to regional subsidence and earthquake effects. This technique involves the controlled removal of small volumes of soil from the sandy silt formation of horizon A beneath the north side of the foundation.

All three approaches have been the subject of intense investigation. Numerical and centrifuge modelling of the north pressing slab have shown that the response of the tower is somewhat uncertain and, if positive, is small while the induced settlements are large. Full-scale trials of the electro-osmosis showed that the ground conditions at Pisa are not suited to this method. Both these methods require costly civil engineering works prior to commencement of the stabilization work. Work on the method of soil extraction is proving much more positive but before describing it a major set-back took place in September 1995 and first this will be described.

A set-back

It is not widely appreciated that the decree establishing the commission has never been ratified. The position in Italian law is that a decree has to be ratified by the Italian Parliament within 2 months of publication or else it fails. Thus, every 2 months, the decree relating to the commission has to be renewed. On a number of occasions the commission has been suspended because of delays in the renewal of the decree. Such an arrangement makes the commission very vulnerable to media and political pressures. Moreover, long-term planning becomes very difficult.

Shortly after the successful application of the temporary counterweight a view emerged that the commission was politically vulnerable and that something needed to be done that would clearly demonstrate the effectiveness of the work so far. There was also considerable concern among some members that, should the commission cease to exist, the unsightly lead counterweight would be left in position for many years. Therefore, a scheme was developed to replace the lead weights with 10 tensioned cables anchored in the lower sands at a depth of about 45 m as shown in Fig. 17. Additional benefits of this proposal were seen to be that the increased lever-arm would give a slightly larger stabilizing moment than the lead counterweight and tensions in the anchors could be adjusted to 'steer' the tower during implementation of induced subsidence. It is important to appreciate that this 10-anchor solution was always intended to be temporary.

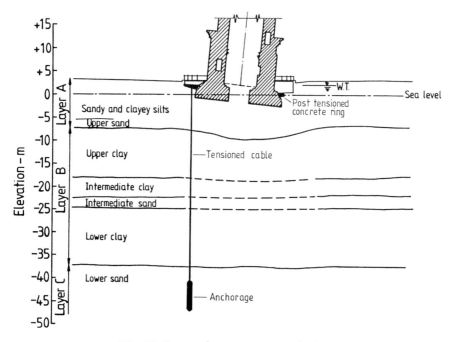

Fig. 15 Ten anchor temporary solution.

The major problem with the 10-anchor solution is that the anchors have to react against a post-tensioned concrete ring around the tower foundation and this involves excavation beneath the catino at the south side—an operation of the utmost delicacy as it is below the water table. Various schemes for controlling the water were considered and it was decided to employ local ground freezing immediately beneath the catino floor but well above foundation level. The post-tensioned concrete ring was to be installed in short lengths so as to limit the length of excavation open at any time.

Shortly before commencement of the freezing operation exploratory drilling through the floor of the catino revealed the existence of an 80 cm thick ancient concrete (conglomerate) layer, which had evidently been placed by Gherardesca in 1838. There are no archaeological records of this conglomerate and its discovery came as a complete surprise. A key question was whether it was connected to the tower. Exploratory drilling was carried out to investigate the interface between the conglomerate and the masonry foundation. A circumferential gap was found all around the foundation and it was concluded that the conglomerate was not connected to the masonry. Work then started on installing the post-tensioned concrete ring.

Freezing commenced on the north side and the northern sections of the ring were successfully installed. The freezing operations consisted of 36 hours of continuous freezing using liquid nitrogen followed by a mainten-ance phase when freezing was carried out for 1 hour/day so as to control the expansion of the ice front. Some worrying southward rotation of the tower did take place during freezing at the north but this was recovered once thawing commenced. Of far greater concern was the discovery of a large number of steel grout-filled pipes connecting the conglomerate to the masonry foundation. These were installed by Girometti in 1934 when the foundation masonry was grouted. In none of the engineering reports of the time is there any reference to these grout pipes or of the conglomerate.

In September 1995 freezing commenced on the south-west and south-east sides of the foundation. During the initial 36 hours of continuous freezing no rotation of the tower was observed. However, as soon as the freezing was stopped for the maintenance phase the tower began to rotate southward at about 4 arc seconds per day. The operation was sus-pended and the southward rotation was controlled by the application of further lead weights on the north side. The resulting southward rotation of the tower was small, being about 7 arc seconds, but the counterweight had to be increased to about 900 t. The main concern was the uncer-tainty about the strength of the structural connection between the con-glomerate and the masonry formed by the steel grout pipes. In view of this uncertainty the freezing operation was abandoned and work on developing the permanent solution was accelerated.

Induced subsidence by soil extraction

Figure 18 shows the proposed scheme whereby small quantities of soil are extracted from layer A below the north side of the tower foundation by means of an inclined drill. The principle of the method is to extract a small volume of soil at a desired location leaving a cavity. The cavity gently closes due to the overburden pressure causing a small surface subsidence. The process is repeated at various chosen locations and very gradually the inclination of the tower is reduced.

Two key questions had to be addressed.

1. Given that the tower is on the point of leaning instability, is there a risk that extraction of small quantities of soil from beneath the north side will cause an *increase* in inclination?

2. Is the extraction of small volumes of ground in a controlled manner feasible, will the cavities close and what is the response at the soil/ foundation interface?

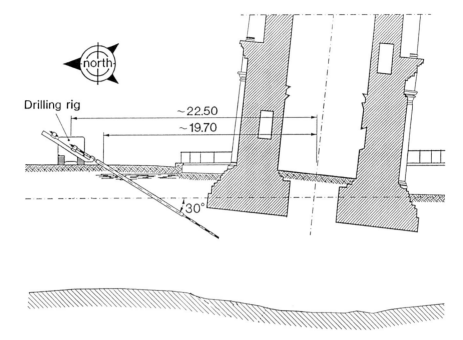

Fig. 16 Induced subsidence by soil extraction.

The first issue has been studied in great detail using two independent approaches—numerical modelling and physical modelling on the centrifuge. The numerical model described previously was used to simulate the extraction of soil from beneath the north side of the foundation. Even though the tower was on the point of falling over it was found that, provided extraction takes place north of a critical line, the response is always positive. Moreover, the changes in contact stress beneath the foundations were small. Advanced physical modelling was carried out on a centrifuge at ISMES in Bergamo. As for the numerical modelling, the ground conditions were carefully reproduced and the model was calibrated to give a reasonably accurate history of inclination. The test results showed that soil extraction always gave a positive response.

The results of the modelling work were sufficiently encouraging to undertake a large-scale development trial of the drilling equipment. For this purpose a 7 m diameter eccentrically loaded instrumented footing was constructed in the piazza north of the baptistry as shown in Fig. 19. The objectives of the trial were:

(1) to develop a suitable method of forming a cavity without disturbing the surrounding ground during drilling;

Fig. 17 Soil extraction trial showing 7 m diameter eccentric-ally loaded footing and inclined drill.

(2) to study the time involved in cavity closure;

(3) to measure the changes in contact stresses and pore water pressures beneath the trial footing;

(4) to evaluate the effectiveness of the method in changing the inclination of the trial footing;

(5) to explore methods of 'steering' the trial footing by adjusting the drilling sequence; and

(6) to study the time effects between and after the operations.

It must be emphasized that the trial footing was not intended to represent a scale model of the tower.

The results of the trial have been very successful. Drilling is carried out using a hollow-stemmed continuous flight auger inside a contra-rotating casing. When the drill is withdrawn to form the cavity an instrumented probe located in the hollow stem is left in place to monitor its closure. A cavity formed in the horizon A material has been found to close smoothly and rapidly. The stress changes beneath the foundation were found to be small. The trial footing was successfully rotated by about 0.25° and directional control was maintained even though the ground conditions were somewhat non-uniform. Rotational response to soil extraction was rapid taking a few hours. Very importantly, an effective system of communication, decision taking, and implementation was developed.

Concluding remarks

Both numerical and physical modelling of the response of the tower to soil extraction has proved positive. The large-scale trials of the drilling technology has shown that the method works for the soil in layer A. A decision has been taken by the commission to carry out preliminary soil extraction beneath the north side of the tower itself. The objective is to observe the response of the tower to a limited and localized intervention. Before this work is undertaken a safeguard structure is to be constructed in the form of a horizontal cable stay attached to the tower at the third storey. If the preliminary soil extraction proves successful it is estimated that it will take about 2 years to reduce the inclination of the tower by about half a degree, which will be barely visible. It is also anticipated that it will be necessary to stabilize the groundwater level in the vicinity of the tower and to carry out some strengthening of the masonry at the south side of the second storey.

At the time of writing the operation of the commission has again been suspended and it is unlikely that work will recommence on the tower before the middle of 1998.

References

1. Ministero dei Lavori Pubblici. (1971). *Ricerche e studi sulla Torre di Pisa ed i fenomeni connessi alle condizione di ambiente.* 3 Vol., I.G.M., Florence.
2. Burland, J.B. and Viggiani, C. (1994). Observazioni sulcomportamento della Torre di Pisa. *Rivista Italiana di Geotecnica*, **28** (3), 179–200.
3. Potts, D.M. and Gens, A. (1984). The effect of the plastic potential in boundary value problems involving plane strain deformation. *International Journal for Numerical and Analytical Methods in Geomechanics*, **8**, 259–86.
4. Schofield, A.N. and Wroth, C.P. (1968). *Critical state soil mechanics.* McGraw-Hill Book Co., London.
5. Burland, J.B. and Potts, D.M. (1994). Development and application of a numerical model for the Leaning Tower of Pisa. *International symposium on pre-failure deformation characteristics of geo-materials*. IS-Hokkaido '94, Japan, Vol. 2, 715–38.
6. Hambly, E.C. (1985). Soil buckling and the leaning instability of tall structures. *The Structural Engineer*, **63A** (3), 77–85.
7. Burland, J.B., Jamiolkowski, M., Lancellotta, R., Leonards, G.A., and Viggiani, C. (1994). Pisa update—behaviour during counterweight application. *ISSMFE News*, **21**, (2).
8. Burland, J.B. (1990). Pisa Tower. A simple temporary scheme to increase the stability of the foundations. *Unpublished report to the Commission*.
9. Terracina, F. (1962). Foundations of the Tower of Pisa. *Geotechnique*, **12** (4), 336–9.

JOHN BURLAND

Born 1936, educated in South Africa and studied civil engineering at the University of Witwatersrand. Returned to England in 1961, working with Ove Arup & Partners, Consulting Engineers. Obtained a PhD at Cambridge University and joined the Building Research Station in 1966, becoming Head of the Geotechnics Division in 1972 and Assistant Director in charge of the Materials and Structures Department in 1979. In 1980 was appointed to the Chair of Soil Mechanics at Imperial College of Science, Technology and Medicine. Was responsible for the design of the underground car park at the Palace of Westminster and the foundations for the Queen Elizabeth II Conference Centre. Is a member of the Italian Prime Minister's Commission for the stabilization of the Tower of Pisa, and of the international Board of consultants advising on the underpinning of the Metropolitan Cathedral of Mexico City. Was London Underground's expert witness for the Parliamentary Select Committees on the Jubilee Line Extension and CrossRail. Was awarded the Kelvin Gold Medal for outstanding contributions to engineering in 1989.

An arts/science interface: medieval manuscripts, pigments, and spectroscopy

Introduction

Much of our cultural heritage is enshrined in manuscripts, paintings, pottery, china, enamels, faience, and other artwork and has been preserved for hundreds and, in some cases, thousands of years. The analysis and identification of pigments on or in such artefacts and the interpretation given to the presence of particular pigments on items from a particular place at a particular date encompass many disciplines spanning both the arts and the sciences. An important role of the scientific disciplines is in the identification of the techniques and materials used in the production of an artefact and the bearing which this may have on our understanding of the development and spread of chemical technology, of artistic styles and techniques, and on the whereabouts of trading routes. Being at the arts/science interface, this subject attracts as much interest from art historians, librarians, and museum scientists as from conservationists and research scientists.

There have been enormous advances in the techniques of analytical chemistry over the past 50 years, to the extent that the identification of any pigment may now be straightforward. However, most techniques currently in use are either intrinsically destructive of a pigment, or cannot be applied *in situ*. Such techniques clearly cannot be used to identify pigments on important manuscripts held throughout the world in galleries and libraries as these operate non-sampling policies. This situation has had the consequence that many of the most significant, beautifully illustrated, and otherwise well-documented artworks currently held in public and private collections have, to date, been poorly characterized in respect of the pigments with which they have been illuminated. Accordingly, a great deal of interesting sociological and technological detail about these works of art has yet to be revealed. It is now considered that Raman microscopy is the best single technique for the identification of pigment grains on manuscripts, paintings, and other artefacts in that it combines the required attributes of being reliable, sensitive, non-destructive, largely immune to interference (from other pigments, binder, and fluorescence), and is applicable to *in situ* studies.[1] Moreover, as the excitation lines usually used lie in the visible region of the spectrum, the spatial resolution possible is ≤ 1 μm. The situation has the consequence that adjacent pigment grains of this size can separately be studied, which is important where the components of a pigment mixture are to be identified.

Despite the above attributes of Raman microscopy its development as an analytical technique for the above purposes has been slow until recently. This has been partly due to the intrinsic weakness of the Raman effect[2,3] in the absence of resonance effects[4] and partly to the moderate detection capabilities (semiconductor or diode), which were available. Many of these problems have, however, been solved with modern spectrometers and many new features, especially as regards CCD (charge-coupled device) detectors, are continually being developed.

The aims of this article are to outline the purposes behind pigment studies, to indicate the nature of the pigments used throughout time on manuscripts, paintings, and other artefacts, to convey an idea as to how the technique works, to review a number of recent applications of great interest, and to point the way to future developments in the area.

Purpose of the examination of pigments

The purposes behind examining pigments from works of art are matters to do with characterization, restoration, conservation, authentication, and dating.[5] Thus, they are:

1. To identify the pigment, its crystal form and its place of origin; and to determine whether it is uniform in composition and particle size at different depths or whether it is layered.

2. To establish whether restoration of damage is feasible, given the identity of the pigments to be replaced.

3. To consider proper measures to conserve a work of art from the effects of heat, light and gaseous pollutants.

4. To consider whether knowledge of the pigments present on a work of art may give an indication as to the date of the work and hence to authenticity. Although minerals are difficult or impossible to date, the year in which a synthetic pigment was first made is usually well established and provides a marker for an indirect form of dating.

The artists' concerns in a pigment are with texture, permanence of colour, compatibility with other pigments, fastness in media, wettability, miscibility, oil absorption, stability of consistency, tinting strength, hiding power, transparency, drying effects, toxicity, and possible presence of adulterants or impurities.[1,7] The analyst, restorer, and authenticator will be concerned to learn whether the pigment identified is one normally found on a work of art executed at the particular time and place to which the artwork is attributed and whether the same pigment is used in other areas of the work.

Pigments—inorganic

Inorganic pigments have the advantage over organic ones of generally being the more stable photochemically. Most of those used before the eighteenth century were minerals, and many required little treatment to produce the required artists' pigment. Nevertheless, certain very long established pigments, e.g. Egyptian blue and lead(II) antimonate, are synthetic and required not inconsiderable chemical knowledge for their synthesis.[5-9] The colours of inorganic pigments arise from ligand-field, charge-transfer, intervalence charge-transfer transitions and/or from specular reflectance, i.e. mirror-like reflectance from the top surfaces of crystallites. As was well known to early artists, the depth of colour of any pigment is related to the size of its particles, as this affects the balance between the diffuse reflectance (controlled by the absorption coefficient and the bandwidth) and the specular reflectance (controlled by complementary factors).[10,11] The fact that deep blue crystals of $CuSO_4.5H_2O$ become much paler when ground to a powder is a well known illustration of this effect.

By way of example, some of the materials which have been used as blue pigments at different periods are listed alphabetically in Table 1 together with, in each case, the chemical formula, the basis of the colour absorption, and (if synthetic) the date of first manufacture. Similar tables are available for black, orange-brown, green, red, white, and yellow pigments and more extensive information can be had from a wide range of books[1,5-9] and pigment company literature. Although the identification

Table 1. Blue inorganic pigments

Pigment	Chemical name	Formula	Date[a]	Transition[b]
azurite	basic copper(II) carbonate	$2CuCO_3.Cu(OH)_2$	min.	LF
cerulean blue	cobalt(II) stannate	$CoO.nSnO_2$	1821	LF
cobalt blue	cobalt(II)-doped alumina glass	$CoO.Al_2O_3$	1775	LF
Egyptian blue (cuprorivaite)	calcium copper(II) silicate	$CaCuSi_4O_{10}$	3rd millenium BC	LF
lazurite (from lapis lazuli)	sulphur radical anions in a sodium aluminosilicate matrix	$Na_8[Al_6Si_6O_{24}]S_n$	min./1828	S_3^- CT
manganese blue	barium manganate(VII) sulphate	$Ba(MnO_4)_2 + BaSO_4$	1907	CT
phthalocyanine blue (Winsor blue)	copper(II) phthalocyanine	$Cu(C_{32}H_{16}N_8)$	1936	π–π^*
posnjakite	basic copper(II) sulphate	$CuSO_4.3Cu(OH)_2.H_2O$	min.	LF
Prussian blue	iron(III) hexacyanoferrate(II)	$Fe_4[Fe(CN)_6]_3.14$–$16H_2O$	1704	IVCT
smalt	cobalt(II) silicate	$CoO.nSiO_2$ $(+ K_2O + Al_2O_3)$	~1500	LF
verdigris	basic copper(II) acetate	$Cu(O_2CCH_3)_2.2Cu(OH)_2$	min.	LF

[a] The pigment is either specified to be a mineral (min.), and/or the date of its first manufacture is listed. Listing taken from ref. 1, which also contains listings of common black, orange-brown, green, red, white and yellow pigments.

[b] LF = ligand field transition; CT = charge transfer transition; IVCT = intervalence charge transfer transition; π–π^* = electric-dipole–allowed charge transfer transition of the phthalocyanine ring system.

of any one pigment will not, of itself, provide precise information as to the date of a piece of artwork, if this information is coupled with that gained by analogous studies on pigments of all other hues, it is then probable that a worthwhile comment on this matter can be made. Clearly, this is of great importance in the valuations of artworks.

Organic pigments and dyes

Among the most common dyes extracted from plants by medieval dyers were indigo from woad for blue, alizarin and madder for red, luteolin from weld for yellow, and crocetin from saffron, also for yellow; all had been established by the Middle Ages, and were widely used on manuscripts. Other organic dyes were extracted from marine molluscs (Tyrian purple, i.e. 6,6′-dibromoindigotin),[12] carmine from scale insects (kermes or cochineal),[8] sepia from cuttlefish, Indian yellow from cow urine,[8] and yet others from many different lichens, etc. Many of the commonly used organic dyes[1] fluoresce and/or are prone to photochemical degradation. An interesting consequence of degradation of one component (say, weld) of a mixture of pigments (e.g. lapis lazuli with weld) is that the colour of the mixture will change with time, posing awkward decisions for conservators as to the extent of restoration which is appropriate.

Raman microscopy

Many different techniques have been used to try to identify pigments or dyes on manuscripts, paintings and other artefacts, notably scanning electron microscopy (SEM), X-ray fluorescence (XRF), X-ray diffraction (XRD), particle-induced X-ray emission (PIXE), particle-induced gamma-ray emission (PIGE), infra-red spectroscopy, ultraviolet/visible spectroscopy, and optical microscopy. Some of these techniques are specific to the elements present, some to the chemical groups present, some to the compounds themselves, and some to the pigment material in bulk. The most recent to be applied is Raman microscopy which is a variant on Raman spectroscopy. The effect, which was first recognized by C.V. Raman in Calcutta in 1928, and which earned the discoverer the Nobel Prize in Physics for 1930, is a consequence of the irradiation of a sample (gaseous, liquid, or solid) with a monochromatic (i.e. single colour) beam of light (nowadays from a laser) and the analysis of the inelastic components of the light scattered by the sample, i.e. the components scattered with frequencies (wavenumbers) different from that of the excitation line.[2,3] The inelastic (the so-called Raman) scattering gives rise to many bands, up to $3N - 6$ of them for a non-linear molecule, where N is the number of atoms in the molecule. These bands—collectively referred to as a Raman spectrum—are highly specific in wavenumber, intensity, and bandwidth to the sample, and thus constitute a unique fingerprint of the latter. Indeed Raman spectroscopy is a remarkably effective analytical technique; it has been argued that, when coupled to a microscope (as in Raman microscopy), it is the ideal technique for the identification of pigment grains on manuscripts, paintings, and other artefacts, for the reasons given in the Introduction. In a typical set-up (Fig. 1), a microscope is coupled to a spectrometer with a sensitive diode array or CCD detector. The laser beam (frequency ν_0) is brought to a focus on each pigment grain in turn by use of the microscope objective ($\times 50$ or $\times 100$). The Raman scattering at frequencies $\nu_0 \pm \nu_i$, where the ν_i are the frequencies of the internal transitions of the scattering species, retraces the path of the incident beam, being collected by the same objective before being directed by a beam splitter to the monochromator and then the detector. The data are then processed and displayed on a screen or as hard copy. Low laser powers (< 1 mW) are used in order to eliminate the possibility of pigment degradation. Many different laser lines of different frequencies can be used in order to obtain the optimum Raman spectrum, including infra-red (1064 nm) excitation in the case of Fourier transform (FT) Raman spectroscopy.

Case studies

Two spectroscopic studies are now outlined to reveal how, in the one case, the identities of the chromophores in a well known pigment were established and, in the other, how research in the nineteenth century in synthetic chemistry in the area of materials science led to the development of a whole range of closely related pigments with colours which could be changed systematically with change of composition.

The intense royal blue colour of the mineral lapis lazuli, has been highly prized for at least 5000 years; its synthetic equivalent, ultramarine blue has been known since 1828. Not only was the semi-precious gem-stone admired as such, but the pigmentary properties of the material have been highly valued since the sixth century, and especially during and after the Middle Ages. The origin of the colour was not understood until a combination of resonance Raman and electron spin resonance studies led to the identification of the chromophores (the parts of the pigment responsible for the light absorption) as sulphur radical anions

157

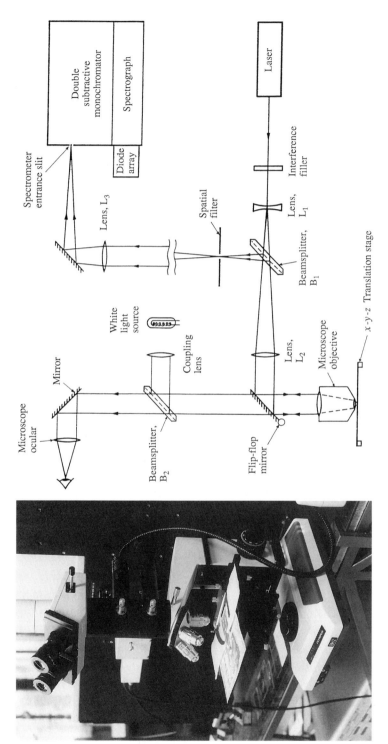

Fig. 1 A Raman microscope together with a schematic representation of the optical cofiguration.

trapped (effectively matrix isolated) within the cubic holes of the host aluminosilicate (sodalite) cage, albeit present in very low ($\leq 1\%$) proportions.[13] The chromophores present were shown to be S_3^- (λ_{max} = 610 nm, ω_1 = 550.3 cm^{-1}) together with some S_2^- (λ_{max} = 380 nm, ω_e = 590.4 cm^{-1}). The key Raman observations on this pigment were the long resonance Raman progressions observed in the totally symmetric stretching mode of each radical anion, features which are characteristic of S_3^- and S_2^- when substituted into alkali halide matrices. Not all pigments, of course, have such intense and highly characteristic Raman spectra as ultramarine blue, but none the less this study illustrates the importance of Raman spectroscopy in characterizing certain minerals and other materials in an easy and highly effective manner.

Cadmium sulphide was first recommended as a yellow artists' pigment in 1818, but the native minerals (greenockite and hawleyite) could not be much used owing to their scarcity. It was not until about 1846 that cadmium sulphide became available commercially, whereupon it soon became popular among impressionist painters, e.g. see Claude Monet's 'Bordighera', 1884, and Charles Demuth's 'Gladioli No. 4', 1925, etc.[8] In addition, it was discovered that selenium could be made to substitute for sulphur, leading to the formation of cadmium selenosulphides with colours ranging from orange through to maroon depending upon the selenium content. This gave the impressionists a wide continuous range of pigment hues to draw upon, and substantially increased the size of their palettes.

Several recent case studies are now discussed in order to illustrate the way in which Raman microscopy has been used to probe the identity of pigments—particularly the inorganic ones—on medieval manuscripts, paintings, and other works of art. Such work has resolved many ambiguities in the minds of art historians in a quick and definitive manner.

Lucka Bible c. 1270

Plate 3 shows the historiated initial letter 'I' from 'In principio …' at the beginning of the book of Genesis in a bible known as the *Lucka Bible*, which is one of many that were made in Paris at c. 1270.[14] Eight pigments have been identified on this bible by Raman microscopy. Six of them were much used (white lead, vermilion, red lead, lapis lazuli, azurite, and orpiment), while two others (realgar and malachite) were found to be present only in trace amounts, possibly as unintended components of orpiment and azurite, respectively. The Raman spectra of the eight pigments (Fig. 2), all inorganic, are distinctive and thus allow clear identification in each case.[14]

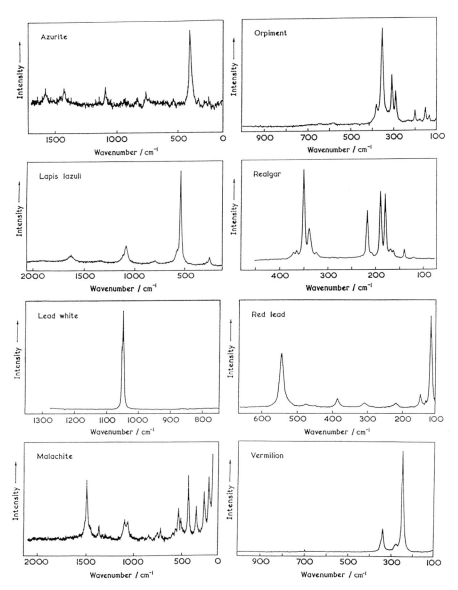

Fig. 2 The Raman spectra of pigments identified on the historiated letter 'I' (see Plate 3).[14]

Skarð copy of the Icelandic Law Book, the Jónsbók c. 1360

Three historiated initials and c. 15 illuminated initials with associated background painting and embellishments from the Skarð copy of the Icelandic Law Book, the *Jónsbók*, have been examined.[15] Six pigments were identified by Raman microscopy, namely vermilion, orpiment, realgar, red ochre, azurite, and bone white. The pigments responsible for some of the green and blue colours could not be identified unambiguously, although the diffuse reflectance spectra of the lighter green and blue pigments appear to arise from verdigris or variants thereof. The dark green colours arise from verdigris, possibly mixed with green earth. Neither red lead nor white lead, pigments commonly used in northern Europe, were found on this manuscript; whether this was due to lack of availability (despite the existence of trading routes, neither pigment being native to Iceland) or to artistic preference is not clear.

German manuscript, sixteenth century

The elaborately historiated initial letter 'R' on a sixteenth century German manuscript has been extensively studied, and eight pigments— azurite, lead tin yellow type I, malachite, vermilion, white lead, red lead, carbon, and massicot—have been identified thereon.[1] It has been shown by Raman microscopy that in this case the two shades of blue on the garments arise from the same pigment, azurite, rather than two different ones, the illuminator having used less azurite and more binder to produce the lighter shade. The deeper blue arises from coarse grains of pigment ($\sim$ 30 μm diameter) whereas the lighter blue arises from fine grains ($\sim$ 3 μm diameter); this effect, as discussed earlier, arises from the different relative importance of diffuse and specular reflectance with differently sized grains.[10,11]

It is of particular interest that the dark grey colour of the pillar top was not obtained by use of a single pigment but by colour subtraction of a mixture of at least the seven pigments, white lead, carbon, azurite, vermilion, red lead, massicot, and lead tin yellow type I. The mixture is evident at ×1000 magnification in Plate 4 and, by Raman microscopy, each different pigment grain (some down to 1 μm across) may be identified. Mixing of pigments was a common practice of artists in order to obtain hues unavailable from a single pigment.

Studies on Persian, Latin, German, and Chinese manuscripts and on a Qur'an

Similar Raman studies have now been completed on a wide range of Persian,[16] Latin,[17] German (Plate 5),[17] and Chinese[18] manuscripts, and on a Qur'an (thirteenth century, from Iran or Central Asia),[19] and have led to the establishment of the palette in each case. By way of illustration, one of the German manuscripts studied—a fifteenth century Book of Prayers

(MS Ger 4) written in Alsatian dialect and held in the DMS Watson Library, University College London—contains many skilfully illuminated initials with rich floral decorations (Plate 5). The pigments used to decorate the initials include vermilion (red), azurite (blue), and white lead (both in its own right and to temper other colours), as well as gold and silver. Unusually, the green pigment present was not malachite, a copper-based pigment, but a mixture of lapis lazuli (blue) and lead tin yellow type I (Pb_2SnO_4); this result could readily be established by Raman microscopy.[17]

Eight pre-tenth century manuscript fragments and one textile fragment from in or near Dunhuang, an oasis town on the Silk Road in Chinese Central Asia, were shown to be illuminated in each case with vermilion, HgS, on the basis of the unique and intense Raman spectrum of this mineral.[18]

Study and identification of degradation products: Byzantine/Syriac Gospel lectionary, thirteenth century

There is great potential for Raman spectroscopy to provide information which would assist conservators to preserve manuscripts and to identify degradation products of pigments. By way of illustration, a study of a very rare, early thirteenth century Byzantine/Syriac Gospel lectionary has been carried out in which unambiguous proof was given as to the nature of a serious conservation problem (the apparent transformation of a white pigment to a black compound) affecting virtually all of the 60 illuminations in this massive volume (Plate 6).[20] Analysis by Raman microscopy proved unambiguously that the white pigment was lead white (basic lead(II) carbonate) and that the black compound was lead(II) sulphide. The possible cause of the degradation of the white pigment was discussed, as were the merits of a proposed treatment with hydrogen peroxide to 'reverse' the process. The artist's palette was also characterized and, in addition to lead white and lead(II) sulphide, a further five pigments were identified unambiguously: vermilion (mercury(II) sulphide), lapis lazuli, orpiment (arsenic(III) sulphide), realgar (arsenic(II) sulphide), and pararealgar. Pararealgar, a yellow pigment, is a light-induced transformation product of the orange/red pigment realgar, As_4S_4, and had not previously been identified on any manuscript.

The thermal degradation of white lead in static and flowing air and in nitrogen and in oxygen atmospheres is very complicated, and has recently been studied by Raman, thermoanalytical and X-ray diffraction techniques in order to define the conditions under which the mineral pigments Pb_3O_4 (orange-red), tetragonal PbO (litharge, yellow-orange) and orthorhombic PbO (massicot, yellow) are formed.[21]

Other degradation studies

Many other studies concerned with degradation and conservation problems are currently being addressed. These include:

(1) the study of the interconversion of HgS (red) to HgS (black) and its
possible reversal;

(2) the study of the photochemical decomposition of orpiment (As_2S_3);

(3) the study of the wavelength dependence of the photochemical con-
version of α- or β-realgar (As_4S_4, both orange) to para-realgar (yellow);

(4) the study of the evolvement of acetic acid from verdigris and basic
verdigris, which leads to changes in the colour from blue-green to
brown-green and possible degradation of the paper, and

(5) the study of malachite/azurite interconversion.

Authenticity studies

The dating aspect of the technique is, of course, of interest to auction
houses such as Sotheby's. As indicated earlier, the information gained
relates solely to whether or not synthetic pigments of known first date of
manufacture are, or are not, present on a manuscript. Clearly, the
identification of a twentieth century pigment on a purportedly much
earlier manuscript would be a cause of great concern—greatly affecting
the value—and several studies of this sort have been carried out. One
currently under discussion is aimed at finally establishing whether or
not the famous Vinland map is authentic. This purportedly pre-
Columbian map of the world includes, significantly, the north-east
coastline of the USA. Forty years after its discovery in a secondhand
bookstore in 1957 it is still unclear whether or not it is a fake.[22] The
difficulty hinges on two scientific investigations: the first, by polarized
light microscopy on 20–30 microsamples removed from the surface of
the map and found in the ink layer, indicated the presence of titanium
dioxide (anatase); this material could not used as a white pigment before
1923, when it was first able to be manufactured pure. The second, by
PIXE (a technique whereby elemental composition can be determined
over sizeable areas of the map but with much poorer (1 mm) spatial resolution) led to no evidence for abnormal levels of titanium. The implied
incompatibility of these results could almost certainly be resolved by
Raman microscopy, taking advantage of the ability with this technique
to range over the whole map at very high (1 μm) resolution.

Pigment sections from paintings

There are several advantages in taking pigment samples for analysis. These are: (i) sampling does not require the removal of an artefact from its permanent location; (ii) well-planned sampling is a one-off operation that needs never to be repeated and so involves much less handling of the artefact; (iii) only the sample, not the entire artefact, is subjected to irradiation; and (iv) samples may be able to be taken in such a way that the lacunae left by excision of the pigment particles are not discernible to the naked eye; they may also be obtainable, in the case of manuscripts, from offsets transferred on to the opposite page. However, although pigment sampling may be permitted from a painting, it is almost never the case from a manuscript owing to the fragility of the latter.

A recent case of interest in which samples from paintings were, indeed, made available for Raman microscopy involved the study and characterization of lead tin yellow. This pigment has a long and complicated history owing in part to the fact, not originally appreciated, that it exists in two distinct forms, type I which is Pb_2SnO_4 and which has a tetragonal structure isostructural with Pb_3O_4, and type II which is $PbSn_{0.76}Si_{0.24}O_3$ and has a defect pyrochlore structure, 24% of the tin sites having been substituted at random by silicon in the lattice.[23] Both types are synthetic and have been in vogue at different periods of time over the past 2000 years. Owing to the similarities in their colours they have sometimes been confused with one another and on other occasions with lead(II) antimonate, $Pb_2Sb_2O_7$ (Naples yellow). The Raman spectra of these pigments are quite different from one another and hence the pigments may readily be distinguished on this basis. A recent Raman study of two late sixteenth century paintings held in the National Gallery has revealed very clearly that the yellow pigment on one of them—'Death of Acteon', Titian, yellow bush in foreground—consists of type I, whereas that on the other—'Allegory of Love IV', Paolo Veronese, man's cloak—consists of type II.[23] Raman studies on samples or cross-sections of pigment layers taken from paintings are comparatively easy to carry out and have widespread analytical application.

Applications to ceramics, pottery, faience, etc.

The technique of Raman microscopy has recently been applied for the first time to the study of pigments on ceramics, pottery, and faience. The first such study was made of the pigments used in the glazes of fragments of medieval items of pottery dating back to the latter half of the thirteenth century and found buried beneath a church in the abandoned village of Castel Fiorentino, near Foggia in southern Italy. This research led for the first time to the identification of lapis lazuli as the blue pigment in a pottery glaze (Plate 7).[24] Brown-black pigments which were also present were identified to be manganese oxides, probably MnO_2.[24]

Another very recent investigation was concerned with the identification of pigments present in ancient Egyptian faience fragments from El-Amarna in the Nile valley (Plate 8). These fragments, which date from the eighteenth dynasty (1350–1334 BC), i.e. from the period of Akhenaten's rule, were discovered by Sir William Flinders Petrie in 1891–1892. The faience pottery, which consists essentially of silica, SiO_2, was found to have been coloured red with red ochre (essentially iron(III) oxide) and yellow with lead(II) antimonate, $Pb_2Sb_2O_7$, on the basis of the unique Raman spectra given by these two inorganic pigments.[25] For reasons not wholly understood, Raman studies of glazed pottery are not successful if the laser beam is directed through the glaze on to the pottery, but only when directed at cross-sections in which the glaze is broken or chipped. So far, red (647.1 nm) excitation appears to be the best to use for Raman studies of pottery.

Future developments

With the development of very much faster detectors (i.e. CCDs), notch filters, and better spectrometer designs, it is now possible to construct Raman spectrometers which need, for their operation, less powerful lasers than previously; the latter may be air cooled and therefore mobile, rather than large, water-cooled, and static. The consequence is that the size of the spectrometer system can now be greatly reduced, which makes it possible to take, if necessary, the complete unit to a gallery or museum for in-house studies rather than the artefact to the laboratory. The optimization of the optical and electronic components of spectrometers for different exciting lines can now also be carried out readily so that potential problems to do with sample heating on irradiation and with sample fluorescence can be overcome. Further important developments involve the use of fibre-optic probes connected to a miniaturized colour video camera in a remote head ('superhead'), which enable *in situ* studies of pigments on wall paintings to be carried out. Moreover, the use of two-dimensional motorized stages to permit both the mapping of surfaces as well as depth profiling *in situ* in order to establish the composition of pigment layers is likely to provide uniquely valuable further information.

Conclusions

Raman microscopy is now clearly established as a major technique for the rapid identification of pigments on manuscripts, paintings, and other artefacts. Its relevance to the identification of the components of pigment mixes at high ($\leq$ 1 μm) spatial resolution is unparalleled. The main difficulties arise with certain organic pigments which either fluoresce (or their supports or binders do), are photosensitive, or fail to yield a Raman spectrum owing to small particle size and/or high degree of dilution, e.g. in a lake. The technique of Raman spectroscopy has, of course, a vast number of technological applications other than to the identification of pigments, namely, to the identification of contaminants in microelectronics, of inhomogeneities formed during the crystallization of polymers, the detection of inclusions in minerals, the monitoring of the curing of polymer resins, and of pesticides in tissues of organisms in the food chain, together with other applications in medicine, jewellery studies, polymer science, and forensic science.[26–28] It should be emphasized, however, that the most effective studies on pigment identification as well as in the other areas of science mentioned are usually carried out by use of Raman microscopy in conjunction with one or more other techniques. In this way the weaknesses of the one may be complemented by the strengths of another.

Acknowledgements

I am most grateful to my recent co-workers in the area who have engendered much interest and displayed great skill in the successful prosecution of this work. These include Drs I.M. Bell, D.A. Ciomartan, and P.J. Gibbs, and L. Burgio, M.L. Curri, M.A.M. Daniels, A. Hardy, and K. Huxley. Support for the work has come from the Leverhulme Trust, the EPSRC, and the ULIRS.

References

1. R.J.H. Clark, *Chem. Soc. Rev.*, 1995, **24**, 187–96.
2. D.A. Long, *Raman spectroscopy*, McGraw-Hill, New York, 1977.
3. R.J.H. Clark and R.E. Hester (ed.), *Advances in spectroscopy*, Wiley, Chichester, Vols 1–26, 1975–98.
4. R.J.H. Clark and T.J. Dines, *Angew. Chem., Internat. Ed. Engl.*, 1985, **25**, 131–58.
5. D.V. Thompson, *The materials and techniques of Medieval painting*, Dover, New York, 1956.

6. R.J. Gettens and G.L. Stout, *Painting materials*, Dover, New York, 1966.
7. K. Wehlte, *The materials and techniques of painting*, Van Nostrand Reinhold Co. Ltd, New York, 1975.
8. R.L. Feller (ed.), *Artists' pigments*, Cambridge University Press, Cambridge, Vol. 1, 1986.
9. A. Roy (ed.), *Artists' pigments*, Oxford University Press, Oxford, Vol. 2, 1993.
10. R.J.H. Clark, *J. Chem. Educ.*, 1964, **41**, 488–92.
11. W.M. Wendlandt and H.G. Hecht, *Reflectance spectroscopy*, Interscience, New York, 1966.
12. R.J.H. Clark, C.J. Cooksey, M.A.M. Daniels, and R. Withnall, *Endeavour*, 1993, **17**, 191–9.
13. R.J.H. Clark, T.J. Dines, and M. Kurmoo, *Inorg. Chem.*, 1983, **22**, 2766–72 and references therein.
14. S.P. Best, R.J.H. Clark, M.A.M. Daniels, and R. Withnall, *Chem. Brit.*, 1993, **29**, 118–22.
15. S.P. Best, R.J.H. Clark, M.A.M. Daniels, C.A. Porter, and R. Withnall, *Studies Conservation*, 1995, **40**, 31–40.
16. D.A. Ciomartan and R.J.H. Clark, *J. Brazil. Chem. Soc.*, 1996, **7**, 395–402.
17. L. Burgio, D.A. Ciomartan, and R.J.H. Clark, *J. Raman Spectrosc.*, 1977, **28**, 79–83; *J. Mol. Struct.*, 1997, **405**, 1–11.
18. R.J.H. Clark, P.J. Gibbs, K.R. Seddon, N.M. Brovenko, and Y.A. Petrosyan, *J. Raman Spectrosc.*, 1997, **28**, 91–4.
19. R.J.H. Clark and K. Huxley, *Science and technology for cultural heritage*, 1996, **5**, 95–101.
20. R.J.H. Clark and P.J. Gibbs, *Chem. Commun.*, 1997, 1003–4. (The manuscript, valued at c. £1 million, was exhibited at the Metropolitan Museum of Art, New York, at an exhibition of Byzantine art in March 1997.)
21. D.A. Ciomartan, R.J.H. Clark, L.J. McDonald, and M. Odlyha, *J. Chem. Soc. Dalton*, 1996, 3639–45.
22. K.M. Towe, *Acc. Chem. Res.*, 1990, **23**, 84–7.
23. R.J.H. Clark, L. Cridland, B.M. Kariuki, K.D.M. Harris, and R. Withnall, *J. Chem. Soc. Dalton*, 1995, 2577–2582.
24. R.J.H. Clark, M.L. Curri, and C. Laganara, *Spectrochim. Acta*, 1997, **53A**, 597–603; R.J.H. Clark, L. Curri, G.S. Henshaw, and C. Laganara, *J. Raman Spectrosc.*, 1997, **28**, 105–9.
25. R.J.H. Clark and P.J. Gibbs, *J. Raman Spectrosc.*, 1997, **28**, 99–103. (The items concerned are held in the Petrie Museum, University College London.)
26. J. Corset, P. Dhamelincourt, and J. Barbillat, *Chem. Brit.*, 1989, 612–16.
27. G. Turrell and J. Corset (ed.), *Raman microscopy*, Academic Press, London, 1996.
28. A. Paipetis, C. Vlattas, and C. Galiotis, *J. Raman Spectrosc.*, 1996, **27**, 519–26.

ROBIN J.H. CLARK

Born in 1935, and educated at the Universities of Canterbury and Otago before being awarded a PhD in Inorganic Chemistry at University College London in 1961. He joined the academic staff at that Institution shortly thereafter, becoming a Professor in 1982, Dean of Science 1988–89, and Sir William Ramsay Professor and Head of Department in 1989. His research in inorganic chemistry and spectroscopy has led to the publication of nearly 400 scientific papers, three books, and 36 edited books. He has held many visiting professorships and has lectured at over 250 universities and institutions in 31 countries throughout the world. He has given the Royal Society of Chemistry's Tilden, Nyholm, Thomas Graham and Harry Hallam Lectures and has served on many national and international committees, including the Royal Society and Royal Institution Councils. He is also Chairman of the Advisory Council and Trustee of the Ramsay Memorial Fellowships Trust. He was elected an Honorary Fellow of the Royal Society of New Zealand in 1989, a Fellow of the Royal Society and a Member of the Academia Europaea in 1990, a Fellow of the Royal Society of Arts in 1992, and a Fellow of University College London in 1993.

THE ROYAL INSTITUTION

The Royal Institution of Great Britain was founded in 1799 by Benjamin Thompson, Count Rumford. It has occupied the same premises for nearly 200 years and, in that time, a truly astounding series of scientific discoveries has been made within its walls. Rumford himself was an early and effective exponent of energy conservation. Thomas Young established the wave theory of light; Humphry Davy isolated the first alkali and alkaline earth metals, and invented the miners' lamp; Tyndall explained the flow of glaciers and was the first to measure the absorption and radiation of heat by gases and vapours; Dewar liquefied hydrogen and gave the world the vacuum flask; all who wished to learn the new science of X-ray crystallography that W.H. Bragg and his son had discovered came to the Royal Institution, while W.L. Bragg, a generation later, promoted the application of the same science to the unravelling of the structure of proteins. In the recent past the research concentrated on photochemistry under the leadership of Professor Sir George (now Lord) Porter, while the current focus of the research work is the exploration of the properties of complex materials.

Towering over all else is the work of Michael Faraday, the London bookbinder who became one of the world's greatest scientists. Faraday's discovery of electromagnetic induction laid the foundation of today's electrical industries. His magnetic laboratory, where many of his most important discoveries were made, was restored in 1972 to the form it was known to have had in 1854. A museum, adjacent to the laboratory, houses a unique collection of original apparatus arranged to illustrate the more important aspects of Faraday's immense contribution to the advancement of science in his fifty years at the Royal Institution.

Why the Royal Institution Is Unique

It provides the only forum in London where non-specialists may meet the leading scientists of our time and hear their latest discoveries explained in everyday language.

It is the only Society that is actively engaged in research, and provides lectures covering all aspects of science and technology, with membership open to all.

It houses the only independent research laboratory in London's West End (and one of the few in Britain)—the Davy Faraday Research Laboratory.

What the Royal Institution Does for Young Scientists

The Royal Institution has an extensive programme of scientific activities designed to inform and inspire young people. This programme includes lectures for primary and secondary school children, sixth form conferences, Computational Science Seminars for sixth-formers and Mathematics Masterclasses for 12–13 year-old children.

What the Royal Institution Offers to its Members

Programmes, each term, of activities including summaries of the Discourses; synopses of the Christmas Lectures and annual Record.

Evening Discourses and an associated exhibition to which guests may be invited.

An annual volume of the *Proceedings of the Royal Institution of Great Britain* containing accounts of Discourses.

Christmas Lectures to which children may be introduced.

Meetings such as the RI Discussion Evenings; Seminars of the Royal Institution Centre for the History of Science and Technology, and other specialist research discussions.

Use of the Libraries and borrowing of the books. The Library is open from 9 a.m. to 9 p.m. on weekdays.

Use of the Conversation Room for social purposes.

Access to the Faraday Laboratory and Museum for themselves and guests. Invitations to debates on matters of current concern, evening parties and lectures marking special scientific occasions.

Royal Institution publications at privileged rates.

Group visits to various scientific, historical, and other institutions of interest.

Evening Discourses

The Evening Discourses have been given regularly since 1826. They cover all aspects of science and technology (with regular ventures into the arts) in a form suitable for the interested layman, and many scientists use them to keep in touch with fields other than their own. An exhibition, on a subject relating to the Discourse, is arranged each evening, and light refreshments are available after the lecture.

Christmas Lectures

Faraday introduced a series of six Christmas Lectures for children in 1826. These are still given annually, but today they reach a much wider audience through television. Titles have included: 'The Languages of Animals' by David Attenborough, 'The Natural History of a Sunbeam' by Sir George Porter, 'The Planets' by Carl Sagan and 'Exploring Music' by Charles Taylor.

The Library

The Royal Institution library reflects the functions and the activities of the RI. The subject coverage is science, its history, its role in society including education, and its interaction with religion, literature, and the arts. The emphasis is on the popular science books, the history of science, and the research monographs of interest to the research group in the Davy Faraday Research Laboratories.

It is probably the only library of its kind specializing in the public understanding of science, that is science for the non-specialist. It also has a junior section.

Schools' Lectures

Extending the policy of bringing science to children, the Royal Institution provides lectures throughout the year for school children of various ages, ranging from primary to sixth-form groups. These lectures, attended by thousands, play a vital part in stimulating an interest in science by means of demonstrations, many of which could not be performed in schools.

Seminars, Masterclasses, and Primary Schools' Lectures

In addition to educational activities within the Royal Institution, there is an expanding external programme of activities which are organized at venues throughout the UK. These include a range of seminars and master classes in the areas of mathematics, technology and, most recently, computational science. Lectures aimed at the 8–10 year-old age group are also an increasing component of our external activities.

Teachers' Workshops

Lectures to younger children are commonly accompanied by workshops for teachers which aim to explain, illustrate, and amplify the scientific principles demonstrated by the lecture.

Membership of the Royal Institution

Member

The Royal Institution welcomes all who are interested in science, no special scientific qualification being required. By becoming a Member of the Royal Institution an individual not only derives a great deal of personal benefit and enjoyment but also the satisfaction of helping to support the unique contribution made to our society by the Royal Institution.

Family Associate Subscriber

A Member may nominate one member of his or her family residing at the same address, and not being under the age of 16 (there is no upper age limit), to be a Family Associate Subscriber. Family Associate Subscribers can attend the Evening Discourses and other lectures, and use the Libraries.

Associate Subscriber

Any person between the ages of 16 and 27 may become an Associate Subscriber. Associate Subscribers can attend the Evening Discourses and other lectures, and use the Libraries.

Junior Associate

Any person between the ages of 11 and 15 may become a Junior Associate. Junior Associates can attend the Christmas Lectures and other functions, and use the Libraries. There are also visits organized during Easter and Summer vacations.

Corporate Subscriber

Companies, firms and other bodies are invited to support the work of the Royal Institution by becoming Corporate Subscribers; such organizations make a very valuable contribution to the income of the Institution and so endorse its value to the community. Two representatives may attend the Evening Discourses and other lectures, and may use the Libraries.

College Corporate Subscriber

Senior educational establishments may become College Corporate Subscribers; this entitles two representatives to attend the Evening Discourses and other lectures, and to use the Libraries.

School Subscriber

Schools and Colleges of Education may become School Subscribers; this entitles two members of staff to attend the Evening Discourses and other lectures, and to use the Libraries.

Membership forms can be obtained from: The Membership Secretary, The Royal Institution, 21 Albemarle Street, London W1X 4BS. Telephone: 0171 409 2992. Fax: 0171 629 3569.